Complete Guide to the Fragment Molecular Orbital Method in GAMESS

From One Atom to a Million, at your Service

Other Title by the Author

The Fragment Molecular Orbital Method: Practical Applications to Large Molecular Systems

Complete Guide to the Fragment Molecular Orbital Method in GAMESS

From One Atom to a Million, at your Service

Dmitri G. Fedorov

National Institute of Advanced Industrial Science and Technology (AIST), Japan

 World Scientific

NEW JERSEY · LONDON · SINGAPORE · BEIJING · SHANGHAI · HONG KONG · TAIPEI · CHENNAI · TOKYO

Published by

World Scientific Publishing Co. Pte. Ltd.

5 Toh Tuck Link, Singapore 596224

USA office: 27 Warren Street, Suite 401-402, Hackensack, NJ 07601

UK office: 57 Shelton Street, Covent Garden, London WC2H 9HE

Library of Congress Cataloging-in-Publication Data
Names: Fedorov, Dmitri, author.
Title: Complete guide to the fragment molecular orbital method in GAMESS :
    from one atom to a million, at your service / Dmitri G. Fedorov,
    National Institute of Advanced Industrial Science and Technology (AIST), Japan.
Description: New Jersey : World Scientific, [2023] | Includes bibliographical references and index.
Identifiers: LCCN 2022031926 | ISBN 9789811263620 (hardcover) |
    ISBN 9789811263637 (ebook for institutions) | ISBN 9789811263644 (ebook for individuals)
Subjects: LCSH: Molecular orbitals. | Computational chemistry.
Classification: LCC QD461 .F363 2023 | DDC 541/.28--dc23/eng20230201
LC record available at https://lccn.loc.gov/2022031926

British Library Cataloguing-in-Publication Data
A catalogue record for this book is available from the British Library.

For any available supplementary material, please visit
https://www.worldscientific.com/worldscibooks/10.1142/13063#t=suppl

Desk Editor: Shaun Tan Yi Jie

Typeset by Stallion Press
Email: enquiries@stallionpress.com

To my mother, who always encouraged and supported me

About the Author

Dr. Dmitri G. Fedorov is Senior Researcher at the Research Center for Computational Design of Advanced Functional Materials (CD-FMat), National Institute of Advanced Industrial Science and Technology (AIST), Japan. He is a leading developer of FMO since 2002. He has published more than 160 peer-reviewed papers and co-edited the book *The Fragment Molecular Orbital Method: Practical Applications to Large Molecular Systems* (2009). His current research interests include development and applications of the FMO method, analysis of molecular interactions, and parallelization of quantum-mechanical methods for high-performance computing. He is a qualified key-punch operator and holds a Ph.D. in Chemistry from Iowa State University, USA.

Contents

About the Author   vii

Chapter 1 Introduction   1

Chapter 2 Getting Started   3

2.1 Practical Summary of FMO   3
   2.1.1 How FMO Works   3
   2.1.2 Energy Decomposition   7
   2.1.3 Methods Interfaced with FMO   11
   2.1.4 Conventions in Notation   12
2.2 Installation of GAMESS   15
2.3 Understanding Parallel Execution   19
   2.3.1 Distributed Data Interface (DDI)   19
   2.3.2 Generalized Distributed Data Interface (GDDI) 20
   2.3.3 Memory Usage   22
2.4 How to Run GAMESS   23
   2.4.1 Running Sequentially on 1 Core   25
   2.4.2 Running on 1 Node in Parallel   26
   2.4.3 Running on Multiple Nodes   28
2.5 General Structure of GAMESS Input Files   30
2.6 FMO Input Files   33
   2.6.1 Basis Set ($DATA) and Atomic Coordinates
       ($FMOXYZ)   34
   2.6.2 Fragmentation and QM Methods ($FMO)   37
      2.6.2.1 Assignment of atoms to fragments 37
      2.6.2.2 General FMO settings   38

2.6.2.3 Charges and multiplicities of fragments   39
2.6.2.4 QM specifications of fragments   39
2.6.2.5 Multiple layers   40
2.6.2.6 Options for individual fragments and segments   40
2.6.2.7 Subsystem definition   41
2.6.2.8 Approximations   42
2.6.2.9 Analytic gradient   47
2.6.2.10 Array dimensions   49
2.6.3 Analyses and Properties ($FMOPRP)   50
2.6.4 Covalent Boundaries Between Fragments ($FMOBND)   53
2.6.5 Hybrid Orbitals for Covalent Boundaries ($FMOHYB)   56
2.6.6 Model Systems in AFO ($AFOMOD)   57
2.7 FMO Result Files (LOG, DAT, TRJ, RST, and F40)   58
2.8 How to Begin Your Own FMO Calculations   59
2.9 Parallelization Strategy   61
2.9.1 The Golden Rule   61
2.9.2 Uneven Task Sizes   62
2.9.3 Load Balancing   63

Chapter 3 Basic FMO Calculations   **67**

3.1 Treating Solvent   67
3.1.1 Explicit QM Solvent   67
3.1.2 Effective Fragment Potential (EFP)   67
3.1.3 Continuum Solvent Models (PCM and SMD)   69
3.2 Desolvation Penalty in Binding (SBA)   74
3.3 Solvent Screening   76
3.4 Polarization, Interaction, and Binding Energies   78
3.4.1 Reference States (0, PL0, PL, and CT)   79
3.4.2 Polarization of Fragments and Subsystems   81
3.4.3 Pair Interactions (PIE)   82
3.4.4 Backbone, Connected and Unconnected Dimers   83
3.5 Analyses for FMO   84
3.5.1 Segments Versus Fragments   84

3.5.2 Segments for Electronic Energy Decomposition 85
3.5.3 Segments for Vibrational Energy Decomposition 87
3.5.4 How to Define Segments 87
3.5.5 Energy Decomposition Analyses for Fragments
 (PIEDA and EDA) 89
  3.5.5.1 Energy decomposition in PIEDA 90
  3.5.5.2 Energy decomposition in EDA3 92
  3.5.5.3 Input files and an application example 92
3.5.6 Partition Analysis for Interactions Between
 Segments (PA) 94
3.5.7 Partition Analysis for Vibrations (PAVE) 97
3.5.8 Subsystem Analysis for Analyzing Binding (SA) 100
3.5.9 Fluctuation Analysis for MD (FA) 107
3.5.10 Free Energy Decomposition Analysis (FEDA) 108
3.6 Periodic Boundary Conditions (PBC) 111
3.7 Geometry Optimizations 113
3.8 Molecular Dynamics (MD) 117
3.8.1 Creating Input Files for MD 117
3.8.2 Processing Results of MD (RDF) 119
3.9 Hessians, IR and Raman Spectra 120
3.9.1 Vibrational Entropy and Enthalpy 122
3.9.2 IR Spectra 123
3.9.3 Raman Spectra 123
3.10 Chemical Reactions 124
3.10.1 Static Reaction Path (IRC) 124
3.10.2 Dynamic Reaction Mapping (MD) 125
3.10.3 Crossing of Energy Surfaces (MEX) 129
3.11 Atomic Charges, Multipole Moments, and Spin
 Populations 130

Chapter 4 Building up Complexity **133**

4.1 Parametrized Methods 133
4.1.1 Density Functional Theory (DFT) 133
4.1.2 Corrections for SCF (DFT-D and HF-3c) 134
4.1.3 Density-functional Tight-binding (DFTB) 135
4.1.4 Interface with Molecular Mechanics (MM) 138

4.1.5 Pseudopotentials   140
    4.1.5.1 Effective Core Potentials (ECP)   141
    4.1.5.2 Model Core Potentials (MCP)   142
4.2 Unrestricted (UHF and UDFT) and Open-shell (ROHF)
Methods   144
4.3 Electron Correlation   145
    4.3.1 Core Inconsistency Problem   146
    4.3.2 Møller-Plesset Perturbation Theory (MP2)   146
    4.3.3 Coupled Cluster (CCSD(T))   148
4.4 Multiconfigurational Self-consistent
Field (MCSCF)   148
4.5 Electronic Excited States   151
    4.5.1 SCF Approaches (ROHF, UHF, and MCSCF)   151
    4.5.2 Single CIS/TDDFT Chromophore Fragment   151
    4.5.3 Multiple TDDFT Chromophore Fragments (FRET) 154
    4.5.4 TDDFT Calculations in Solution (PCM)   155
4.6 Multiple Layers   156
4.7 Frozen Domain (FD)   157
4.8 Molecular Orbitals and Their Energies (LCMO)   159
4.9 Properties on a Grid   161
    4.9.1 Total Properties   163
    4.9.2 Fragment Properties   164
4.10 Struggling with Convergence   164
    4.10.1 Charge Instability   165
    4.10.2 Basic HF/DFT Convergers   167
    4.10.3 Fine-tuning Convergers   168
    4.10.4 DFTB Convergers   168
    4.10.5 MCSCF Convergers   169
    4.10.6 Alternating QM Methods   169
    4.10.7 Mixing Convergers   170
    4.10.8 Convergence Criteria   171
4.11 Fragmentation   172
    4.11.1 Hybrid Orbital Projection (HOP)   175
    4.11.2 Adaptive Frozen Orbitals (AFO)   183
    4.11.3 BDA Corrections for PIEs   185
    4.11.4 Molecular Clusters   188
    4.11.5 Atomic Ions   188

4.11.6 Polymers (Including Polypeptides,
        Polysaccharides, and Polynucleotides)   189
4.11.7 The Junction Rule   189
4.11.8 Inorganic Systems (Nanomaterials)   190
4.11.9 Periodic Systems (Liquid and Solid State)   192

Chapter 5 Advanced Techniques   **193**

5.1 How to Compute Polarization Energies   193
     5.1.1 Stepwise Polarization in PIEDA   193
     5.1.2 Polarization in AP   197
     5.1.3 Polarization in Solution (IEA)   198
5.2 Embedding Types   200
5.3 Diffuse Basis Sets   204
5.4 Dual Basis Sets (AP)   205
5.5 Restarting Jobs   206
     5.5.1 Geometry Restarts   206
     5.5.2 MD Restarts   206
     5.5.3 Fragment-Specific Text Restarts   207
     5.5.4 All-Fragment Binary Restarts   207
5.6 Using FMO for Non-FMO Tasks   208
5.7 Temperature, Entropy, and Free Energy   209
5.8 Acceleration Tricks   210
     5.8.1 Acceleration of I/O   210
     5.8.2 Acceleration of Data Processing   212
     5.8.3 Acceleration via Scientific Means   214
5.9 How to Calculate Millions of Atoms   215
5.10 Improving GDDI Performance   219
     5.10.1 Analyzing Parallel Performance   220
     5.10.2 Toll of Data Servers   223
     5.10.3 Three-Level GDDI   224
     5.10.4 How to Optimize Parallel Efficiency   225
             5.10.4.1 Similar fragment sizes   226
             5.10.4.2 Uneven fragment sizes   227
             5.10.4.3 Multiple layers   231
             5.10.4.4 Protein-ligand complex   232
     5.10.5 File-Less GAMESS Execution   233

Chapter 6 Reference Materials **235**

  6.1 Conclusions and Outlook   235
  6.2 Troubleshooting   236
      6.2.1 Failure to Start   238
      6.2.2 External Abnormal Termination   240
      6.2.3 Internal Abnormal Termination   240
      6.2.4 Scientific Failure   243
      6.2.5 Input Errors   244
      6.2.6 Version Change   247
  6.3 Collection of Sample Input and Output Files   248
      6.3.1 FMO2-RHF/STO-3G, HOP   250
      6.3.2 FMO2-DFT/3-21G, AFO, LCMOX   252
      6.3.3 FMO2-UHF/6-31G*/D3(BJ)   254
      6.3.4 FMO3-CCSD(T)/AP/cc-pVDZ:aug-cc-pVDZ   257
      6.3.5 FMO2-MCSCF/6-31G   259
      6.3.6 FMO2-DFTB3/SMD, PA   262
      6.3.7 FMO2-MP2/PCM, PIEDA + SA   267
      6.3.8 FMO3-SCS-MP2/MCP, EDA3   271
      6.3.9 FMO2-CAM-B3LYP/6-311G*, Density + MEP   272
      6.3.10 FMO-TDDFT/4-31G, FRET   275
      6.3.11 FMO2-HF3c Optimization   277
      6.3.12 FMO2-RHF/STO-3G:B3LYP/6-31G*/FDD,
            TS Search   279
      6.3.13 FMO2-LC-BOP/ 6-31G, PAVE   283
      6.3.14 FMO2-DFTB3/PBC, NVT US MD, FA   285
      6.3.15 Lego Input Maker   287
  6.4 Processing Results   287
  6.5 GAMElish Dictionary   289
  6.6 Suggestions for Further Reading   291

Instructions for Accessing Online Supplementary Material   293
References   295
Index   297

CHAPTER 1

Introduction

Ich am y-hote Escalibore, Unto a king fair tresore.

— *Specimens of Early English Metrical Romances,*
G. Ellis

This book is written to assist an interested user to run fragment molecular orbital (FMO) calculations as implemented in the GAMESS program.[1,2] A succinct guide to installation of GAMESS is provided, as well as some minimal general information on the GAMESS input file structure. A short introduction to FMO and a summary of its features in GAMESS are also included.

The author thanks all persons who contributed to the development of FMO in GAMESS, Prof. Kazuo Kitaura for many illuminating discussions regarding FMO, Dr. Michael Schmidt for invaluable help with GAMESS development, Dr. Sarom Leang for assistance with the updating process, and Prof. Mark Gordon for heading the GAMESS project over many years. The author is indebted to Dr. Vladimir Sladek for making many useful suggestions for improving this book.

The task undertaken by the author is to describe the rich variety of FMO features developed by contributors to GAMESS. The book is written in a complete manner for a beginner, but intermediate and advanced features are also covered. Few very technical and complicated features are omitted, for example, the use of specialized hardware (such as GPU). Brief installation instructions are given only for a typical UNIX-based PC cluster.

All descriptions of GAMESS features are limited to FMO. Any statement that such-and-such calculation is possible or impossible applies exclusively to FMO unless stated otherwise.

This book describes the program usage; both lengthy mathematical formulations and citations of numerous methodological papers are avoided throughout, as it would obfuscate the narrative. Input and output files are for the September 30, 2021 R2 Patch 2 version of GAMESS, released in April 2022. Any development of FMO in local versions of GAMESS, that is not available in the distributed code, is excluded from this book.

The author has to apologize for infesting this book with words that may appear as meaningless discourse markers: "usually", "normally", "in most cases", etc. In fact, every such phrase was inserted after a painful deliberation. The abysmal complexity of options makes any general statement nearly impossible to formulate, because there may be an obscure case that makes the statement false. If the author were to explicitly list such clauses, this book would be unreadable; on the other hand, completely forsaking these qualifiers would make the book a collection of statements that are false in certain cases. Thus, upon encountering such qualifiers the reader should realize that in a rare combination of options a description might not be true.

A disclaimer has to be made: although the author took utmost efforts to provide the information in the book as accurately as possible, either due to an omission on the part of the author or due to a change in the program, the behavior of GAMESS may differ from that described in this book. For the former, the author humbly apologizes, and the latter is an inevitable outcome of progress. In any case, the author cannot accept responsibility for a discrepancy between the book and the program.

CHAPTER 2

Getting Started

2.1 Practical Summary of FMO

The idea to accelerate quantum-mechanical (QM) calculations by using locality of interactions in terms of structural subunits (fragments) has assumed various forms over many decades.[3] There is a variety of fragment-based methods, many of which are similar to each other in physical principles, but differ in some details, important or trivial. FMO is one out of perhaps several dozens of such methods.

In addition to FMO, GAMESS has several other low-scaling approaches: effective fragment potential (EFP), elongation method (ELG), divide-and-conquer (DC), and effective fragment orbital method (EFMO). FMO is also available in ABINIT-MP[4] and PAICS[5] programs, but this book focuses exclusively on FMO in GAMESS.

The development of FMO has been described in several reviews[6] and book chapters (Section 6.6). One review[7] and one book chapter[8] have dealt with the development of FMO in GAMESS, and these two sources are recommended for those interested in getting references for various FMO methods, with detailed mathematical derivations, omitted as much as possible from this book.

2.1.1 How FMO Works

It is important for practical applications to understand all steps in an FMO calculation, for example, to deal with divergence of self-consistent field (SCF) calculations. The description of the main concepts and equations is rather succinct, to serve as a reference.

In FMO, a molecular system is divided into N fragments (monomers). For example, $(H_2O)_3$ can be divided into three H_2O fragments. Each atom is assigned to one fragment as a rule, except that one atom per covalent boundary is placed in two fragments (Figure 1). No hydrogen caps are used at boundaries. Instead, each fragment is immersed into an embedding potential of the whole system, rather like the mean-field potential in unfragmented QM calculations.

$$CH_3-H_2C-CH_2-CH_2-CH_2-CH_3$$

Figure 1.   Division of *n*-hexane into three fragments.

The overall scheme of FMO is shown in Figure 2. The embedding potential (often called the electrostatic potential, ESP) is obtained from properties of fragments, typically, as the Coulomb potential from their electron densities and nuclei (Section 5.2). After calculating each fragment, the embedding potential is updated, and fragment calculations are repeated again ("monomers in embedding loop"), until self-consistency is reached and the energy of fragments does not change. In this step **1**, sometimes called monomer SCF, a fully polarized state of fragments is obtained.

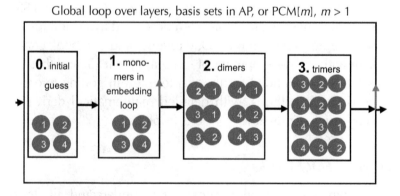

Figure 2.   Computational scheme for a single-point FMO3 calculations of 4 fragments. Each fragment is denoted by a circle. Branching arrows indicate possible iterations in loops.

Step **1** yields a converged embedding, and an internal energy E'_I of each monomer (internal refers to subtracting the contribution of the embedding from converged SCF energies of monomers as a post-processing), and the interaction energy ΔE_I^{solv} of fragment I with continuum solvent (for calculations in vacuum, $\Delta E_I^{solv} = 0$) with the following estimate E^{FMO1} of the total energy,

$$E^{FMO1} = \sum_{I=1}^{N} \left(E'_I + \Delta E_I^{solv} \right) \tag{1}$$

In order to account for interactions between fragments, fragment pairs (also called dimers) are computed. In step **2**, the embedding for each dimer is computed using the electron densities of monomers (each dimer is calculated only once). A pair interaction energy (PIE) ΔE_{IJ} is obtained for each pair of fragments, I and J. These pair interactions are added to the energy of FMO1,

$$E^{FMO2} = E^{FMO1} + \sum_{I=1}^{N} \sum_{J=1}^{I-1} \Delta E_{IJ} \tag{2}$$

ΔE_{IJ} is the QM interaction between two fragments, which can be decomposed into various contributions, such as electrostatic or dispersion (Section 3.5.5). If all interactions in nature were two-body, then E^{FMO2} would recover the exact energy of the full system without fragmentation. It does indeed happen for $N = 2$. However, there are higher-body interactions in larger systems.

Consider charge transfer (CT) in three fragments I, J, and K, for example, in a chain of 3 water molecules $I...J...K$ bound by two hydrogen bonds $I...J$ and $J...K$. The CT between I and J is coupled to the CT between J and K. If some charge of J is transferred to I, then charge-deprived J' may be less willing to give charge to K. This coupling is quite small, as can be visualized by plotting electron density differences. The extra three-body coupling is just a small fraction of the two-body CT.

E^{FMO2} is a good approximation of the exact energy E, whereas E^{FMO1}, especially for systems with covalent boundaries between fragments, is not, and the total energy E^{FMO1} is never used in lieu of E, although FMO1 as a method has some usage for computing polarization energies, electron density, or electronic excitations.

For a better estimate of E, it is possible to compute triples of fragments (also called trimers) in step **3**, and obtain E^{FMO3} by adding three-body contributions ΔE_{IJK} to E^{FMO2},

$$E^{FMO3} = E^{FMO2} + \sum_{I=1}^{N}\sum_{J=1}^{I-1}\sum_{K=1}^{J-1}\Delta E_{IJK} \tag{3}$$

ΔE_{IJ} and ΔE_{IJK} are symmetric with respect to a permutation of indices, so that sums go over unique combinations. Many-body expansions (MBE) in Eqs. (2) and (3) explicitly involve n-body terms, for $n = 2$ and 3, respectively; however, there is also an explicit coupling of the electronic state of n-mers to the embedding potential, so the actual degree of the inclusion of many-body effects is $n + 1$.

There is a global loop (Figure 2) over monomers, dimers, and trimers for multiple layers (loop over layers), auxiliary polarization (AP, with a loop over basis sets), and polarizable continuum model (PCM) using a many-body solute potential, PCM[m], $m > 1$ (loop over the PCM embedding). Solvation model density (SMD) is in most ways very similar to PCM, and a likewise loop is also done for SMD.

Because there is a coupling of QM interactions, any truncation of E^{FMOn} for $n < N$ is an approximation. In GAMESS, $n = 3$ (FMO3) is the highest available level.

Fortunately, high n-body couplings for $n > 2$ are short-ranged (whereas the electrostatic component of the two-body interaction is long-ranged). For attaining good computational efficiency, there are approximations for a rapid estimation of ΔE_{IJ}, when two fragments I and J are far separated, whereas ΔE_{IJK} can be neglected if three fragments are far separated (decoupled many-body effects cancel out). Therefore, the sums in Eqs. (2) and (3) involve a limited number of QM dimers and trimers, whose total number is linear in N (except for small N).

This is in essence the computational scheme of FMOn. It is for the user to decide whether to employ FMO2 or FMO3, although in a vast majority of applications, FMO2 is used.

Table 1. Steps in an FMO*n* calculation.[a]

Index	ID	When used	What is done
	0	always	Fragmentation bookkeeping
1	1	always	Initial guess
1	2	always	Monomer calculations
2	4	$n > 1$	Dimer calculations
3	9	$n > 2$	Trimer calculations
4	2	$n \geq 1$	Correlated monomers[b]
5	6	$n > 1$, RESDIM≠0	ES dimers
6	2	MCSCF	RHF monomers[c]
7	4	MCSCF	RHF dimers[c]

[a]Index *i* is used in step-specific options like NGRFMO(*i*). ID is printed in abnormal terminations.
[b]Index 4 is used for the last iteration in the monomer loop, where correlation or excitation energies are computed.
[c]In FMO2-MCSCF, RHF monomers and dimers are also computed. For MCSCF monomer and dimers, indices of 1 and 2 are used, respectively.

An FMO calculation for a given geometry consists of a sequence of steps listed in Table 1 (it is more detailed than Figure 2). Index of an FMO step is used to set options for different stages in FMO calculations, for example, for index equal to 4, an option NGRFMO(4) can be set to define the number of parallel groups used in dimer calculations, as explained below. On the other hand, step ID is printed in an abnormal termination and may be helpful in dealing with problems, especially when a meager printout level is used and it is not obvious where the calculation stopped.

2.1.2 *Energy Decomposition*

The pair interaction energy in general is computed as

$$\Delta E_{IJ} = \Delta E'_{IJ} + \Delta E^V_{IJ} + \Delta E^{solv}_{IJ} \tag{4}$$

where ΔE^{solv}_{IJ} is the solvent screening. The internal energy contribution to PIE is computed as the difference of the internal energies of dimer *IJ* and two monomers *I* and *J*,

$$\Delta E'_{IJ} = E'_{IJ} - E'_I - E'_J \tag{5}$$

ΔE^V_{IJ} is the explicit energy of CT between fragments I and J in the embedding potential $\mathbf{V}^{IJ}$, whereas the implicit energy of CT is included in $\Delta E'_{IJ}$. In all methods with the exception of DFTB,

$$\Delta E^V_{IJ} = Tr\left(\Delta \mathbf{D}^{IJ} \mathbf{V}^{IJ}\right) \tag{6}$$

where $\Delta \mathbf{D}^{IJ}$ is the difference in the electron density matrices of dimer IJ and monomers I and J. In DFTB, the density is replaced by atomic populations.

In FMO3, there are three-body interactions,

$$\Delta E_{IJK} = \Delta E'_{IJK} + \Delta E^V_{IJK} + \Delta E^{solv}_{IJK} \tag{7}$$

In some variations of FMO, there are several sets of ΔE_{IJ} and ΔE_{IJK}. In AP (Section 5.4), they correspond to different embeddings and basis sets. There is also a combined value computed from them, so the user should be careful in reading such output files.

The expression in Eq. (2) defines the ground state energy. FMO can be used for an excited state i. In such calculations, one chromophore fragment denoted by L is selected for calculating electron excitations. Excited states are also calculated for dimers and trimers including L. The energy of an electronically excited state i is defined by adding the ground state and excitation energies,

$$E^{FMO3,i} = E^{FMO3} + \omega^{FMO3,i} \tag{8}$$

and the total excitation energy is

$$\omega^{FMO3,i} = \omega^i_L + \sum_{\substack{I=1 \\ I \neq L}}^{N} \Delta\omega^i_{IL} + \sum_{\substack{I=1 \\ I \neq L}}^{N} \sum_{\substack{J=1 \\ J \neq L}}^{I-1} \Delta\omega^i_{IJL} \tag{9}$$

where ω^i_L is the excitation energy of state i in fragment L. $\Delta\omega^i_{IL}$ and $\Delta\omega^i_{IJL}$ are the two- and three-body interactions of exciton i in chromophore L with other fragments, respectively, describing delocalization of the electron excitation beyond L. A two-body equation for $\omega^{\text{FMO2},i}$ can also be defined by neglecting three-body terms in Eq. (9). For completeness, $\omega^{\text{FMO1},i} = \omega_L^i$.

A very useful quantity is the amount of CT $\Delta Q_{I \to J}$ from fragment I to J, which is antisymmetric (that is, $\Delta Q_{I \to J} = -\Delta Q_{J \to I}$). A large inter-fragment CT can lead to an FMO accuracy loss. A three-body value $\Delta Q_{I \to JK}$ is computed in FMO3, showing the extra CT from I to JK, in addition to pairwise $\Delta Q_{I \to J}$ and $\Delta Q_{I \to K}$. The three-body CT in trimer IJK can be divided into $\Delta Q_{I \to JK}$, $\Delta Q_{J \to IK}$, and $\Delta Q_{K \to IJ}$.

Because of the charge conservation,

$$\Delta Q_{I \to J} + \Delta Q_{J \to I} = 0 \tag{10}$$

$$\Delta Q_{I \to JK} + \Delta Q_{J \to IK} + \Delta Q_{K \to IJ} = 0 \tag{11}$$

The total solvent-related energy in PCM (SMD) is divided into the solvent-related energy ΔE_I^{solv} of monomers and solvent screening in dimers $\Delta E_{IJ}^{\text{solv}}$ and trimers $\Delta E_{IJK}^{\text{solv}}$,

$$E^{\text{solv}} = \sum_{I=1}^{N} \Delta E_I^{\text{solv}} + \sum_{I=1}^{N} \sum_{J=1}^{I-1} \Delta E_{IJ}^{\text{solv}} + \sum_{I=1}^{N} \sum_{J=1}^{I-1} \sum_{K=1}^{J-1} \Delta E_{IJK}^{\text{solv}} \tag{12}$$

The solvent-related energies are not the same as solvation energies in the conventional sense. The latter refer to the energy gain or loss in the transition from vacuum to solvated state (solute-solvent binding energies), whereas solvent-related energies are solute-solvent interactions. The difference between solvation and solvent-related energies is the solvent-induced solute polarization. Solvation energies can be easily computed in FMO by doing two runs, in solution and in vacuum, and subtracting the total energies.

There are two ways of separating solvent-related energies into monomer and dimer terms in Eq. (12), referred to as the local and partial screenings (Section 3.3). In PCM, the solvent-related fragment energy is

$$\Delta E_I^{\text{solv}} = \Delta E_I^{\text{cav}} + \Delta E_I^{\text{es}} + \Delta E_I^{\text{disp}} + \Delta E_I^{\text{rep}} \tag{13}$$

The solvent screening is the solvent-related dimer term,

$$\Delta E_{IJ}^{solv} = \Delta E_{IJ}^{CT \cdot es} + \Delta E_{IJ}^{es} + \Delta E_{IJ}^{disp} + \Delta E_{IJ}^{rep} \tag{14}$$

where CT·es denotes the interaction of charge transfer in the solute (CT) with the electrostatic embedding of the solvent (es). The CT·es term is usually small, and the screening is dominated by the es term.

In SMD, there is only one non-electrostatic term, the combined cavitation, dispersion, and solvent structure (cds) term, and it corresponds to the cavitation (cav), dispersion (disp), and repulsion (rep) terms in PCM (cdr for short). The dispersion here is between solute and solvent. The es term appears in both PCM and SMD. The solvent structure (SMD) and cavitation (PCM) terms are related to the solvent entropy loss in solvation.

Atomic charges Q_A and multipole moments (such as dipoles $\mathbf{d}$) can be decomposed in a many-body expression similarly to the energy. Dipole moments $\mathbf{d}^I$ of individual fragments are useful for discussing the embedding effects. For a general multipole tensor $\boldsymbol{\mu}$ ($\boldsymbol{\mu}$ can be the atomic charge vector, dipole moment vector, quadrupole matrix, etc.),

$$\boldsymbol{\mu} = \sum_{I=1}^{N} \boldsymbol{\mu}^I + \sum_{I=1}^{N} \sum_{J=1}^{I-1} \Delta \boldsymbol{\mu}^{IJ} + \sum_{I=1}^{N} \sum_{J=1}^{I-1} \sum_{K=1}^{J-1} \Delta \boldsymbol{\mu}^{IJK} \tag{15}$$

For a complex of a macromolecule A and ligand(s) B, it may be useful to add PIEs over fragments I in the macromolecule A, resulting in the total interaction energy (TIE) between A and B,

$$\Delta E_B^{TIE} = \sum_{I \in A} \Delta E_{IB} \tag{16}$$

Sometimes it is convenient (albeit artificial) to compress the complexity of many-body terms by summing over one index. The partial energy of fragment I is

$$E_I^{part} = E_I' + \Delta E_I^{solv} + \frac{1}{2} \sum_{J \neq I} \Delta E_{IJ} \tag{17}$$

The contracted PIE for a pair I,J incorporates three-body interactions involving I and J,

$$\Delta\tilde{E}_{IJ} = \Delta E_{IJ} + \frac{1}{3}\sum_{K \neq I,J} \Delta E_{IJK} \tag{18}$$

The fractions in Eqs. (17) and (18) are used because each dimer would otherwise be counted twice (in IJ and JI) in the partial energies and thrice for PIEs (in IJ, IK, and JK). The following contracted expansions can be defined,

$$E^{FMO2} = \sum_{I=1}^{N} E_I^{part} \tag{19}$$

$$E^{FMO3} = \sum_{I=1}^{N} E_I' + \sum_{I>J} \Delta\tilde{E}_{IJ} \tag{20}$$

2.1.3 *Methods Interfaced with FMO*

A summary of wave functions available for FMO in GAMESS is shown in Table 2, whereas possible RUNTYP tasks (selected in $CONTRL) are summarized in Table 3. Prefixes R, RO, and U mean restricted closed-shell, restricted open-shell, and unrestricted, respectively (Section 4.2). They can be attached to other methods, for example, to Hartree-Fock (HF), resulting in restricted HF (RHF) method. HF is used to denote all variations (RHF, ROHF, or UHF); density functional theory (DFT) encompasses RDFT or UDFT; coupled cluster (CC) can be RCC or ROCC. CIS can only be restricted.

Table 2. Summary of wave functions and methods, grouped by the highest analytic derivatives.

Derivatives	Methods[a]
Analytic Hessians	DFTB, HF/PCM, DFT/PCM
Exact gradients	DFTB/PCM, RMP2/PCM
Approximate gradients	MCSCF/PCM, TDDFT, ROMP2
Energy[b]	UMP2, UTDDFT, TDDFT/PCM, CIS, CC/PCM

[a]PCM is indicated where it can be combined (SMD can be used interchangeably with PCM).

Table 3.   RUNTYP tasks in GAMESS possible with FMO.

RUNTYP	MXATM[a]	Section	Usage
ENERGY	No	3.5	Analyses (e.g., PIEDA, PA, SA).
GRADIENT	No		Check a molecular geometry.
OPTIMIZE (OPTFMO)	Yes (No)	3.7	Energy minimization (geometry optimization).
GLOBOP	Yes	3.1	Global minimum search for EFP fragments.
SADPOINT	Yes	3.10.1	Transition state search.
MEX	Yes	3.10.3	Minimum energy crossing of two spin state surfaces.
MD	No	3.8	Molecular dynamics,
		3.5.9	Fluctuation analysis,
		3.5.10	Free energy decomposition analysis.
IRC	Yes	3.10.1	Mapping reaction profile.
HESSIAN (FMOHESS)	Yes (No)	3.9	Harmonic frequencies,
		3.9.1	Thermochemistry,
		3.9.2	IR intensities and Raman activities,
		3.5.7	Partition analysis of the vibrational energy.
RAMAN	Yes	3.9	Raman activities.
FMO0	No	5.1	Isolated state (no embedding).

[a]This column shows whether MXATM = 2000 limits the total number of atoms in FMO or not.

The value of MXATM cannot be changed in the input; instead, the user can change it in the source file source/mx_limits.src, and recompile GAMESS. In practice the default value of 2000 is enough for FMO, because it applies mainly to the size of individual fragments, and only in rare usages to the whole system.

2.1.4 *Conventions in Notation*

In some analyses described in this book, segments are used. A segment is a set of atoms; see Section 3.5.1 for an explanation of why segments are useful and how they differ from fragments.

Fragments are numbered by capital indices I, J, and segments by small indices i, j. A shorthand notation in the sum "$I < J$" means a double sum, the first sum over I (from 1 to N, the number of fragments), the second sum over J, from I to $J-1$.

Residue fragment names include a dash, as in Asp-10, indicating that they are different from conventional residues (Section 4.11.6). When segments are chosen as conventional residues, their names have no dash, as in Asp10.

In various analyses, capital letters are commonly used to denote solute terms, for example, ΔE_{IJ}^{ES} denotes the electrostatic interaction between two solute fragments I and J. Small letters describe solute-solvent interactions, for example, ΔE_{IJ}^{es} denotes the electrostatic solvent screening in a pair of fragments IJ. There are two dispersion terms (DI for solute-solute and disp for solute-solvent) and two repulsion terms (REP for solute-solute and rep for solute-solvent).

X is used to denote a general n-mer, so that X can be a monomer, a dimer, or a trimer. In some publications, single (E_X') and double (E_X'') primed energies are defined in PCM, with the relation between them $E_X'' = E_X' - \Delta E_X^{solv}$, so that E_X' includes, and E_X'' excludes, the solvent-related energy ΔE_X^{solv}. In this book, E_X' is uniformly used for the "bare" energy of X both in vacuum and solution, that is, E_X' is the solute-only energy excluding ΔE_X^{solv} (as E_X'' in other publications).

$\Delta E_{I(J)}^{A}$ denotes the interaction of solute fragment I with solvent immersing fragment J ($A = $ es, cav, disp, and rep). For example, $\Delta E_{I(J)}^{es}$ is the electrostatic interaction of the electron density of fragment I with the solvent charges induced by fragment J. In this book, for simplicity, $\Delta E_{I(I)}^{A}$ is denoted by ΔE_{I}^{A}.

In tensors, fragment indices are superscripts (as in $\mathbf{D}^I$), leaving space for subscripts of tensor indices ($D_{\mu\nu}^I$), whereas for scalars fragment indices are usually subscripts (E_I).

In multilayer runs, layers are listed in increasing level of complexity separated by a colon. For example, FMO2-RHF/STO-3G:PBE/6-31G* has RHF/STO-3G in the lower layer and PBE/6-31G* in the higher layer.

The names of GAMESS-related files are given relative to the root GAMESS directory (GMSPATH), unless otherwise specified. When providing examples of file names related to FMO calculations, $job is used as a generic name, to be replaced by the actual job name (for example, when running a job exam37.inp, file $job.F30.000 means exam37.F30.000). The extension 000 here means the compute process with the lowest rank 0; if there are more than 1000 cores, the extension has more digits, for example, .00000 is used for running on 65536 cores. The record number of digits in the rank extension for an FMO calculation so far is 6.

In this book, the file that contains the results is called the output file. As described in detail below, in FMO, groups of compute processes can be used together to calculate fragments. Every group writes one such file. Normally, rungms redirects output from group 0 to a log file, so that "output file" in this book normally means $job. log. Other processes with rank *i* write output files $job.F06.*i*. Likewise, group 0 writes a DAT file $job.dat and other processes with rank *i* write $job.F07.*i*. A summary is shown in Table 4.

Non-master processes normally do not write anything but in case of an abnormal termination they may write some error messages to their own output files. It may be noted that while a calculation is

Table 4.   Files produced in a parallel run using 2 groups, with 2 CPU cores each.

Group	0		1	
Rank	0	1	2	3
Status	master		master	
Files	$job.log[a]	($job.F06.001[a]	$job.F06.002[a]	($job.F06.003[a]
	$job.dat[b]	$job.F07.001[b])	$job.F07.002[b]	$job.F07.003[b])
	$job.rst.000[c]			
	$job.trj.000[c]			

[a]$job.log is the main output file, containing messages from rungms and FMO results obtained by group 0. $job.F06.002 contains FMO results of group 1. F06 and F07 files on non-master compute processes are usually empty and they may be suppressed by setting PAROUT = .F. in $GDDI.

[b]$job.dat may contain MOs of fragments done by group 0, data on a grid, and a Hessian (while a job is running, $job.dat is called $job.F07.000, renamed in rungms at the end). $job.F07.002 may contain MOs of fragments done by group 1.

[c]Restart (RST) and trajectory (TRJ) files in MD.

running, the main output file ($job.log) is normally written to the same directory as the input file, whereas all other files are in the run-rime directory SCR (usually, on a local disk of each node), copied by rungms to the restart directory USERSCR after the calculation finishes.

A queuing system may write its own output and error files, that can contain something useful if a failure occurred. When a calculation finishes normally, $job.log and $job.dat contain all the main results; but in case of an abnormal termination, all output and error files may have to be looked into.

2.2 Installation of GAMESS

The installation of GAMESS is described in Table 5. If several compilers are available, for example, gfortran and ifort, one can try to build two separate versions and compare performance. In the author's experience, the difference is not substantial though.

In this book the script-based way of installing GAMESS is described. Alternatively, one can use the "make" route of compiling GAMESS (see README.md), not described here. Some differences exist in the functionality of the script and make ways of compiling GAMESS.

For a math library, 3 scenarios are possible: (1) no external math library, (2) an external BLAS3 library, and (3) MKL. The LAPACK capability of MKL can be used in GAMESS (to use MKL, it may be necessary to set up MKLROOT to the location of MKL prior to compiling GAMESS). For FMO, a BLAS library does not have a major impact on performance in most types of tasks, except for DFTB and some correlated methods like coupled cluster.

The most interesting choice the user has to make in compiling GAMESS is the parallel model, of which there are three main choices: sockets, MPI, and hybrid MPI/OpenMP. Of these three, the last one is of use in a limited set of tasks, for which, however, it is the only way to do them. Namely, to use some methods based on resolution of the identity, an MPI/OpenMP build is required. It is possible to

Table 5.   Steps for installing GAMESS on a Unix cluster.

Step	Description	Comments
1	Prerequisites.	C and FORTRAN compilers, (t) csh shell.
2	Download optional packages.	a. BLAS library (MKL, etc.), b. internal packages (GAMESS-adapted Tinker[a]), c. external packages (libXC, NBO, VeraChem, etc).
3	Apply for a license and download.	Visit GAMESS homepage.[9] tar xzvf gamess-current.tar.gz ; cd gamess
4	Do configuration.	./config
5	Compile parallel library.	cd ddi ; ./compddi[b] mv ddikick.x ..[c] ; cd ..
6	Compile QM code and link.	./compall[b] ./lked[b]
7	Create run-time directory.	Create a run-time directory on a local disk of each compute node (not needed for queuing systems, which create it for each job) and define the directory name as SCR in rungms.
8	Create restart directory.	Create a restart directory in the home directory; set it as USERSCR in rungms.
9	Customize rungms.	Set GAMESS directory as GMSPATH; if needed, define MPI and queuing system details.
10	Queuing system.	For a queuing system (DQS, etc.), create a script to execute rungms.
11	Parallel setup.	Enable password-less scp and ssh to each compute node.[d]

[a]Adapted Tinker can be downloaded from the same place where GAMESS is distributed.
[b]For a copy of the log, the lengthy output can be redirected when running these scripts.
[c]Required only for socket DDI.
[d]Not required for some types of runs.

create 2 executables, a socket one for common tasks and another one with MPI/OpenMP for specialized tasks. Most FMO calculations are not properly parallelized with the hybrid OpenMP/MPI model, thus the main choice is between sockets and MPI (Figure 3).

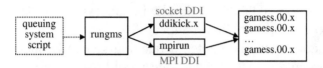

Figure 3. Sequence of the GAMESS ignition process. A script (dashed box) is used in queuing systems. Some MPIs use a running program other than mpirun.

The socket library in GAMESS uses ssh to connect to compute nodes, which can be done both on low-end (Gigabit and its more efficient versions such as 10GbE) and high-end (Infiniband) networks. The latter can be done with IPoIB. For both sockets and MPI, scp and ssh are used to process remote files if multiple nodes are used.

One thing to bear in mind is that rungms is shipped ready to use for sockets. On the other hand, for setting up MPI runs, rungms may need modifications, and the user should be able to do some csh-scripting.

A general advice is: if you have your own cluster, use sockets. On a public cluster, users may be forced to employ MPI due to imposed access restrictions. In MPI, there is a polling problem of data servers (*vide infra*), so the socket model may be faster than MPI (some MPI libraries appear to have dealt with this problem). Which of the two is more efficient depends on the software (MPI library and OS), hardware (CPU and networking devices), and the calculation type. Usage of groups in parallelizing FMO greatly decreases the cost of communications via their localization, and the decision of sockets vs. MPI is not as clear as for flat non-FMO runs, where the power of MPI may be more pronounced.

For a run-time directory SCR defined in rungms, /scr/$user directory is often used (where $user is the user name). For this, /scr should be created (mounted) on each compute node, and made accessible to all users, so that they can create their directories in it. Some public clusters may have decided for you that you do not need local disks. In this lamentable case, SCR can be created on a network disk. On the contrary, USERSCR is typically created on a network disk, in the home directory (USERSCR should be accessible to all compute nodes, so it should not be on a local disk).

At a computational center, a queuing system is often installed, and to execute GAMESS there, one would use a command of that queuing system, with a small script, which sets up details of the execution such as the number of requested nodes and a queue name. On a private cluster, one can log in a node and execute a job interactively, directly by running "rungms" from a command prompt.

To abstract from the low level (MPI or sockets), GAMESS calls wrapper-like subroutines that form distributed data interface (DDI).[10] In flat DDI, there is one "team" of CPU cores. It is also possible to use generalized DDI (GDDI) with multiple teams (called groups). The same standard executable can be used for both DDI and GDDI runs, and the choice between them is made with NGROUP = 0 of $GDDI for DDI, and a positive integer for GDDI. For FMO the general advice is to use GDDI in all runs.

In some cases, a GAMESS executable may be linked to use dynamically loaded .so files, which are a part of Unix. These dynamic libraries should be available on all compute nodes, otherwise GAMESS will not be able to run.

When GAMESS runs, it accesses some files that are stored in the GMSPATH directory. The files relevant for FMO are: (a) ericfmt.dat for computing integrals over Gaussian primitives, (b) MCP files (containing potentials and basis sets), and (c) DFTB parameters. These files, a part of a standard GAMESS distribution, are usually put on a network disk so that compute nodes can access them.

A useful script to know about is gms-files.csh. It defines the location of various external files used by GAMESS, as pertinent to MCP, DFTB, RI-MP2, and other methods. The default locations are sensible so that the script is rarely edited.

It is important to have all tests in tests/standard executed for each installation of GAMESS. The provided script checktst can be run to verify the correctness of all tests. By running the provided set of tests, it can be ensured that the program is correctly built. Better spend an hour now than be puzzled over strange results for months later.

2.3 Understanding Parallel Execution

2.3.1 *Distributed Data Interface (DDI)*

The focus in this section is on the socket version. When a single node is used, a compute process per CPU core is executed (Figure 4) by ddikick.x. Communications between processes are done via System V shared memory.

When more than one node is used, the execution of GAMESS increases in complexity (Figure 5). Instead of one, now two processes per core are run. For example, to execute GAMESS on 2 nodes, each with 3 cores, ddikick.x will run 12 processes. One half of them does calculations and another half serves data to compute processes, so these processes are called data servers. The same executable, like a stem cell, is cloned for all 12 processes, out of which 6 grow to be compute processes and 6 become data servers (the executable has code for both, and the choice is made based on the rank). Any compute process can contact any data server, which happens in sharing large matrices.

Data servers are in a waiting mode, expecting incoming requests for a global sum of some array, or a slice of a distributed matrix, etc.

Figure 4. Execution on 1 node using 3 cores. Compute processes exchange data via shared memory.

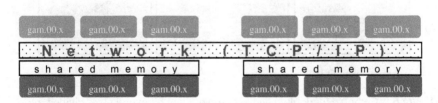

Figure 5. Execution model on 2 nodes, each using 3 cores (the executable name is truncated to gam.00.x). Dark blue compute processes do calculations, light blue processes (data servers) do communications.

While waiting, data servers do not consume CPU time when the socket version is used. For MPI, data servers may use CPU, so only one half of CPU cores may be useful for computations (due to the cost of polling). With hyperthreading, the cost of data servers may be small, but a queuing system may enforce a limit to the allowed number of processes per node, halving the performance.

The user can monitor the workload in real time using Unix commands (such as "top"), and it is possible to get timings from the log file.

2.3.2 *Generalized Distributed Data Interface (GDDI)*

GDDI is used to divide all CPU resources into groups (Figure 6), and to perform individual monomer, dimer, and trimer calculations on separate groups of compute processes. It is optional — one can do FMO calculations without GDDI, by using DDI. When FMO is used with DDI, all CPU cores work together to compute fragments one by one. It is seldom that this way is efficient, and some functionality in FMO cannot be used without GDDI. Therefore, it is recommended that GDDI be always used.

The compute process with the lowest rank in each group is its master, the data server for this process is referred to as the master data server. In Figure 6, there are 3 master compute processes doing calculations and three master data servers. The compute process with

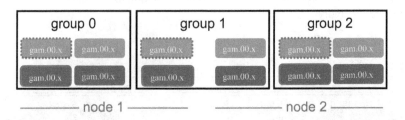

Figure 6.   Compute (dark blue) and data server (light blue) processes divided into 3 groups on 2 nodes of 3 cores each (to do this, 2 physical nodes should be split into 6 logical nodes of 1 core each). Red boxes denote master processes (data servers are dashed). This group definition blatantly defies the node topology for group 1 (transgressing over two nodes), a possible source of performance penalty.

the lowest rank among all groups is called the grandmaster (only 1 per run), and there is a grandmaster data process. Masters and grandmasters work as other processes, and they also do some extra management work unique to them.

GDDI with 1 group is not equivalent to DDI. Think of it as a bicycle (DDI) vs. a bus (GDDI). Both can take you to the same destination, but they are not equivalent: there may be bus lanes and bicycle lanes.

GDDI has many levers that the user should learn, as a pilot learns to fly an airplane. A miserable performance can result if the user blindly relies on the default setup. When using GDDI, it is necessary to set the number of groups. To gain a better parallel efficiency, this can be done for each FMO step (Table 1) separately. There are two ways to set the number of groups. The simple way, that is often reasonable, is to set a uniform NGROUP in $GDDI to be used in all steps.

A more complex way is to define an array of group sizes NGRFMO in $FMOPRP for each FMO step. How does a combination of NGROUP in $GDDI and NGRFMO work? When a GAMESS calculation begins, before FMO is properly started, GDDI performs the division of all available nodes into NGROUP groups. If NGROUP is 0, then GDDI is not used (instead, DDI is employed).

At FMO step *i*, if a value of NGRFMO(*i*) is zero, then no group change occurs, that is, whatever the number of group was, remains to be. For instance if the user defines NGROUP = 4 and NGRFMO(2) = 8, then in the beginning 4 groups are created. Because NGRFMO(1) is not set, it is 0 by default, so monomers (step **1**) are computed with 4 groups (no group change), but for dimers (step **2**), 8 groups are used.

A simple recommendation is: either set only NGROUP to an appropriate value, and define no NGRFMO, or set NGROUP = 1 and define NGRFMO (advised for PCM). Likewise, MANNOD matching NGROUP in $GDDI can be used in the very beginning of a run to do a manual node size definition, whereas MANNOD in $FMOPRP can be used for each FMO step matching NGRFMO.

There is a performance penalty to change the number of groups, so instead of NGRFMO(1) = 2,2, it is better to use $GDDI NGROUP = 2 $END.

2.3.3 *Memory Usage*

Memory is allocated in GAMESS in three different ways, depending on how a particular code is written: (1) replicated memory pool, (2) distributed memory pool, and (3) system pool. The replicated and distributed memory pools are allocated as fixed buffers when GAMESS starts and then memory is internally doled out as needed.

The replicated amount is specified with MWORDS and the distributed amount with MEMDDI, both in $SYSTEM. The units are megawords (MW), 1 MW = 8 MB. The MWORDS amount of replicated memory is allocated on each CPU core, so for a run on 2 nodes, 20 cores per node with MWWORS = 100, 100 × 20 × 8 = 16000 MB is allocated per node. The distributed memory MEMDDI is shared by all cores in all nodes, so that for MEMDDI = 100, 100 × 8 = 800 MB is allocated on 2 nodes, 400 MB per node.

In general, a GAMESS calculation needs some memory of each type. Most FMO jobs can be executed with MWORDS = 100 and MEMDDI = 100, which for an 8 core node is 6400 + 800 = 7200 MB. To learn how much memory is needed, one can in principle attempt to add EXETYP = CHECK to $CONTRL, but for FMO this execution mode is not fully functional (but it can still be useful). For a simple advice, try the "standard" MWORDS = 100 MEMDDI = 100. If either type of memory is insufficient, an error message will be printed.

By changing input options it is possible to reduce the required memory. For example, there are several MP2 codes in GAMESS, all of which produce the same result, but with different memory requirements.

The third type of memory is a run-time allocation ("malloc" or "shmget" in C and "ALLOCATE" in FORTRAN). Some arrays are dynamically allocated from free memory available to OS on the fly as GAMESS runs. These on-the-fly arrays may be very large for parallel CCSD(T), but for other runs they are usually small.

Some FMO tasks can ask for large memory. These are: Hessians (replicated memory) and coupled cluster (system memory). If there are many thousands of fragments, storing data for these fragments takes a fair amount of replicated memory (some hints to reduce this amount are given in Section 5.9). A large amount of replicated memory may be needed for a large fragment. A substantial memory may be needed for storing electron density on a fine grid.

In GDDI, each group gets its share of distributed memory MEMDDI (proportional to the number of ranks). The use of distributed memory is tricky in GDDI, because some arrays are allocated for all groups together (for example, an array for storing file F40 in memory), and others are allocated in a group for a fragment (for example, for CODE = DDI for MP2). For the latter usage, if a GDDI calculation reports that MEMDDI is insufficient, the number of groups can be decreased so that each group gets a larger share.

The replicated memory in MWORDS is used in two ways: (1) genuinely replicated arrays and (2) shared arrays. For the former, the same amount of memory is allocated per core no matter how large a group is; for the latter, the required amount can be reduced if there are more compute processes. If a calculation reports that MWORDS memory is insufficient, one can try to increase the group size, but it may not help (in which case one should reduce the number of compute processes per node; for example, instead of using 24 processes per node, use 8, tripling the memory available per rank).

Other ways to reduce the required memory include choosing a different CODE in $MP2, giving up CC and switching to MP2, using RI methods, or changing the fragmentation (splitting a large fragment). Some acceleration options (e.g., MODGAM = 13 in $DFTB or NINTIC in $INTGRL) may allocate large memory, and they can be removed.

2.4 How to Run GAMESS

Here, some introductory guidance is given for running in UNIX interactively (executing rungms from a command prompt). Among other possibilities, it can be mentioned briefly that a Windows executable is also distributed (which may be somewhat outdated). It is usually executed with a provided batch file, different from rungms. In most cases, only a single node can be used; and it may be advantageous to add NSUBGR = −1 to $GDDI to define multiple teams, similarly to the use of this option described below for rungms.

To run GAMESS, the rungms script is executed with an input file as an argument. As described above, rungms has to be manually edited before it can be executed. The minimal necessary action is to define run-time SCR and restart USERSCR directories in rungms, and create these directories, the former on every compute node, the latter in the home directory. The socket version of DDI requires that "rungms" be executed from the first node (master node) in the list of nodes, after connecting to it with ssh.

The functionality of the socket and MPI models is quite similar, as far as GDDI is concerned. Here, instructions are given mainly for the socket version of DDI. Minor modifications in the arguments of rungms can be done to use it with MPI (in particular for arguments 3–5 that define the details of nodes). For MPI, a node file cannot be passed as an argument to rungms and instead, it should be built inside rungms, possibly using nodes assigned by the queuing system.

A fairly common harmless mistake is misunderstanding the option PARALL in $SYSTEM. This obscure option is for developers, and end users should not set it. Whether or not PARALL is set to .TRUE. or .FALSE has nothing to do with turning parallelization on or off. There is no way to stop parallelization.

For a DDI run, rungms copies the input file to SCR of the master node (but not to other nodes). For GDDI, however, each group should have access to a copy of the input file, so rungms copies the input file to SCR of each node with scp. At the end of a calculation, important result files (DAT, TRJ, RST, etc.) for group 0 are copied back from SCR of the master node to USERSCR. All files that are not copied are deleted from SCR by rungms, including F40, output, and DAT files from all other groups, which can be useful. It may be necessary to edit rungms to save files before their deletion.

A word of caution should be said about stopping GAMESS calculations. It is usually a bad idea to use a queuing system command for it, unless there is no other way. A better approach is to terminate a single GAMESS process (gamess.00.x) with the "kill" command, which should let the parallel manager (ddikick.x, mpirun, etc.) gracefully stop the calculation, and the script rungms can process files. There may be very undesirable consequences from an ungraceful killing, such as that semaphores, used for shared memory, fail to be returned. System resources can become unavailable in a consequent run.

2.4.1 *Running Sequentially on 1 Core*

The simplest is to run on 1 CPU core (known as a sequential run). It is done with

./rungms exam01 00 > & exam01.log

Here, it is assumed that the input file exam01.inp is found in the current directory. The 2nd argument (00) is the version of GAMESS. By default, config builds version 00.

Not all GAMESS tasks can be executed in parallel. Some QM methods in GAMESS cannot use more than 1 core. Some of these are included in the standard tests ("tests/standard"), so all of these tests should be executed on 1 core.

There is another very important aspect of running rungms. When it is executed, it checks if some results from a job with the same name are already present in the restart directory USERSCR, or run-time directory SCR. These two directories usually differ from the directory where the input file is located. The following files are meant: .dat file (called punch file, containing various data for restarts, such as MOs, Hessian, etc.), RST file (MD restart), TRJ file (MD trajectory), etc. If any of these files are found, rungms will stop. The user should either remove all restart files in USERSCR before running the job or modify rungms to delete them automatically. Files in SCR are usually deleted automatically, but may survive if a job is killed ungracefully.

Adding " > &exam01.log" (which may be shell-dependent) redirects the output to the file specified. The results printed by the main process 0 are thus placed in this file.

2.4.2 *Running on 1 Node in Parallel*

For FMO, there are 4 ways of using 1 node in parallel, summarized in Table 6. The concept of a team is used in it, and a team denotes a set of compute processes doing an *n*-mer calculation. In regular GDDI, a team is identically the same as a group. For DDI, there are no groups, but there is 1 team. For an intranode split, 1 GDDI group is divided into multiple teams using NSUBGR = −1. Intranode splits are mainly used for socket DDI, although they can be used for MPI too. However, in MPI there is a different, general way of doing an intranode split, accomplished with the 5th argument of rungms, which defines the size of a logical node (to use it, see the Gnb mode of execution in Table 6).

Fragments can be run nearly independently on teams (groups) of CPU cores, reducing parallel overhead and communications. An example of a parallel run is illustrated in Figure 7, where the setup on the left describes both G1a and G1b, and on the right Gnb. Gna (not shown) would have 5 teams of 1 core each. Other than the team count of 5 in Gna vs. 4 groups in Gnb, the execution scheme of Gna is similar to Gnb in Figure 7.

Although 4 ways are listed for completeness in Table 6, FMO users are discouraged from using G1a in general, because it does not employ GDDI. The difference between G1a and G1b may not be

Table 6.   Different ways to use 1 node with 8 CPU cores in parallel.

Method	NGROUP[a]	Intranode split[b]	Teams	Arguments of rungms[c]
G1a	0	No	1	$job 00 8
G1b	1	No	1	$job 00 8
Gna	1	Yes	8	$job 00 8
Gnb	4	No	4	$job 00 8 1 2[d]

[a]NGROUP is an option in $GDDI defining the number of groups in GDDI. When it is 0, DDI is used.
[b]Accomplished with NSUBGR = −1 in $GDDI.
[c]Usually executed with a redirection, as in "rungms $job 00 8 > & $job.log" (shell-dependent).
[d]This way of executing rungms can be used on 1 node only for sockets, and on any number of nodes for MPI.

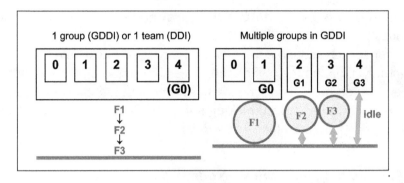

Figure 7. 5 CPU cores (ranks 0...4), shown as boxes, can be used as a single team G0 (left) or divided into 4 groups G0...G3 (right). Fragment tasks are labeled F1... F3. On the left, they are done one by one consequently by G0; on the right, they are done in parallel by the 4 groups. 1 group (G3) has nothing to do (a planning mistake). A larger group G0 may be appropriate if F1 is a larger fragment than F2 and F3 (groups are numbered from 0, and fragments from 1, due to linguistic differences). The red bar indicates a synchronization point (barrier) where all groups meet and exchange information. Green pointed arrows show synchronization loss (idle waiting) for groups G1–G3.

apparent to the user, but there are many minute side effects that arise. Among the two, G1b is preferred. Gna is mainly used for DFTB. For FMO, the most common way among the 4 listed is Gnb.

In Gna, an internal splitting of 1 GDDI group is done, so that each compute process can be used to compute a different n-mer, as if it were a group with 1 member (this complication is needed to overcome the limitation in GDDI that a node may not be divided into groups). For Gna, all nodes are split into teams of 1 compute process each.

For Gnb, as an example, 4 groups are used, into which 8 cores are divided (2 cores per group). It is not necessary to have a divisible number. For 3 groups on 8 cores, the groups will have 3, 3, and 2 cores. The last group is always the smallest. For Gnb, the 5th argument of rungms, 2, means that 2 CPU cores form a logical node (the 4th argument, 1, is a placeholder; it is relevant to MPI but not used for sockets). In this example, rungms defines 4 logical nodes of 2 cores each (because of the 5th argument set to 2), so that it is possible to define up to 4 groups. In GDDI, logical nodes are divided among groups, not cores.

For sockets in Gnb, ssh commands are used to execute GAMESS remotely. Gna, on the other hand, does not use ssh on 1 node (but it needs ssh on more than 1 node). In some cases, ssh may not be able to connect (as in Cygwin) making it impossible to use Gnb, but Gna will work. There are other subtle differences.

2.4.3 *Running on Multiple Nodes*

With sockets, one has to create a node file and pass it to rungms. A node file contains a list of compute nodes and the number of cores on each node. The first node in the node list is called the master node. For instance, if there are 3 compute nodes, called abacus1 (8 cores), abacus2 (12 cores), and abacus3 (16 cores) then the node file "nodelist" can be:

```
abacus1 8
abacus2 12
abacus3 16
```

To execute GAMESS, log in to the master node (abacus1), change directory to where exam01.inp is, and issue

./rungms exam01 00 nodelist > & exam01.log

If the user forgets to ssh to abacus1 and tries to execute rungms from some other node, GAMESS will not start (for sockets).

In the above, rungms uses 8, 12, and 16 cores on those nodes, but each of the nodes may have more cores than that. It is possible to run multiple GAMESS jobs on the same node at the same time, if jobs have different input file names.

What are the rules for co-using nodes? They all must be able to run the same GAMESS executable, and ssh and scp between nodes should be allowed (with interactive login with sockets). It is possible to mix nodes with different CPUs (provided that they can run the same executable). The number of cores and memory on each node may be different. The same user name must be defined on all of the nodes, and a run-time directory specified in SCR should exist on each node.

There is an important limitation of socket GDDI, that the number of groups may not exceed the number of nodes in the node list (although formally the same applies to MPI, there is a way to circumvent the problem). In the above example with 3 abaci, the number of groups may not exceed 3, although in aggregate there are 36 cores. What if you want to have more? In this case, for sockets a node file can be created with redundant entries (for MPI, a similar subdivision can be achieved with the 5th argument to rungms). For example, by splitting a 12-core abacus2 and a 16-core abacus3, a node file with 5 lines can be created.

```
abacus1 8
abacus2 6
abacus2 6
abacus3 8
abacus3 8
```

With this node file, up to 5 groups may be defined in GDDI. A physical node abacus2 with 12 cores is here split into 2 logical nodes of 6 cores each. This is what rungms does internally with its 5th argument: it creates a node list with redundant entries. But rungms can only do it for a single node for sockets. The parallel setup can be checked by finding the following line in the log file (*processors* here mean CPU cores and *nodes* are logical nodes):

PARALLEL VERSION RUNNING ON   12 PROCESSORS IN   1 NODES.

A group can include multiple nodes. Using the above example of 5 lines in nodelist, one can combine these 5 logical nodes into 2 groups by using NGROUP = 2. In this case, the first 3 nodes (abacus1, abacus2, and abacus2) form one group, and the other two (abacus3 and abacus3) form another. It does not make much sense to split a node and then combine it again, with a loss of efficiency. This example is given simply to illustrate the point of how nodes can be combined into groups.

DDI has its limits: the maximum number of nodes (MAXNODES) and cores per node (MAXCPUS). The limits are defined in ddi/compddi (by default, MAXCPUS = 128 and MAXNODES = 32). After changing these values, run compddi, move ddikick.x up if using sockets, and then run lked (Table 5).

2.5 General Structure of GAMESS Input Files

A GAMESS input file is a text file, divided into groups (also called decks, implying playing GAMESS with decks of input cards). All groups may be roughly divided into two categories, unformatted and formatted (in the sense of formats used in FORTRAN). Any group begins with a dollared name (e.g., $CONTRL) and ends with an $END statement. Inside most unformatted group, various keywords are given as $A = B$ (a keyword A is assigned a value B). Some unformatted groups do not use an $A = B$ assignment (such as $DATA). There is a limit of 6 characters per keyword (A) and 8 characters per text values of B. Logical values may be entered either as .TRUE./.FALSE. or as .T./.F.. Floating point numbers can be entered as 1.2e-3 for 1.2×10^{-3}.

There is a convention (but not a strict rule) that options follow the FORTRAN usage: variables starting from I-N are integer, and double precision options start with other letters, whereas for logical variables there is no convention. There is no single/double precision distinction: all floating point values are read in as double precision variables.

To enter an array, the following form is used: $A(I) = B_1, B_2, \ldots, B_m$, where the elements of an array A starting from I are assigned values $B_1 \ldots B_m$. For separating $B_1 \ldots B_m$, either spaces or commas may be used. The element numbering starts from 1. A matrix is entered as a supervector according to the FORTRAN convention, that is, first column, second column, etc. (in an input, the supervector form $A(I) = \ldots$ is used).

Formatted groups also begin with a name (e.g., $VEC) and end with an $END, but anything inside is written in a specific position-related format (such as positions from 1–10 are used for the first number, 11–20 for the second, etc.), which must be strictly observed, and no keywords are specified ($A = B$ is not used). Typically, these groups contain arrays, which are taken from a preceding GAMESS calculation, so that the user does not normally type these numbers. An example is a $VEC group containing molecular orbitals (MO).

Each line in the input has a limit of 80 characters. A dollar symbol (which might be displayed differently in certain language environments) cannot be the first in a line (to prevent it, a space symbol denoted by "⎵" is added, as in "⎵$VEC"). A group can be split into multiple lines simply by continuing onto the next line. Anything between recognized input groups is ignored (treated as a comment). Any line longer than 80 characters is read up to column 80 and anything beyond is silently ignored (no space left on the punch card). It is a treacherous feature because a user may be looking at the input and not noticing that a part of the input is ignored because it goes beyond the 80th column (novice users may like to set their terminal to be 80 characters wide, for an easier spotting of involuntary transgressions).

Most text in the input are converted internally to uppercase, so that both $CONTRL and $contrl are acceptable. The user should refrain from using special symbols such as tabs. Spaces may be used freely in unformatted groups, but some empty lines are mandatory.

The above rules are not always effective. They apply to most input groups but not to all. For some, a customized reading is done, in which case the rules may be different (an example is $QUANPO). In general, the user should assume that the above rules do hold, unless specifically instructed otherwise.

Most keywords have a default value, used if nothing is assigned explicitly. This is why arrays may be given starting from an element other than 1 (if the array has predefined values). Thus, an input file may be very short because most values are predefined.

Some keywords, however, have no predefined values. If no value is specified for such a keyword, GAMESS will abort. For example, a basis set must be specified.

For finding default values, the manual (docs-input.txt) can be consulted. For FMO, many groups are used in a common way with the rest of GAMESS: $SCF, $MP2, $DFT, $DFTB, $BASIS, $STATPT, $SYSTEM, $CONTRL, etc. A few groups (most importantly, $DATA) are used differently, and some groups are FMO-specific ($FMO, etc.).

Some input options are packed. An example of a packed option is MODPAR in $PCM. A packed option is a single integer number. The total value of this option, which the user puts in an input file, should be made by adding several values that are a power of 2. For example, adding $1 + 8 + 256 = 265$ and setting MODPAR = 265 can be done to compact a suboption denoted by $2^0 = 1$, a suboption denoted by $2^3 = 8$, and another $2^8 = 256$ into a single value of MODPAR (the reason for using the powers of 2 is the convenience of unpacking). A single value of 265 carries three different suboptions inside.

Typically, some finetuning options of secondary importance are made compact in this way. To unpack a value, use the binary representation. For example, for 265, it is 1,0000,1001 that clearly shows that three suboptions are set appearing as "1". The powers of 2 they are associated with are their positions counting from the right starting from 0: 0, 3, and 8. This breaks the value into $2^0 + 2^3 + 2^8 = 1 + 8 + 256$.

If two different features related to the same packed option are found in this book, for example, NPRINT = 1 in $FMOPRP in one place discussing output reduction and NPRINT = 8 in another place discussing printing properties, then to use them both in the same run, they should be combined with a logical OR, resulting in NPRINT = 9. If they are put sequentially, as $FMOPRP NPRINT = 1 NPRINT = 8, then the last one is effective and the rest is ignored, i.e., the result is NPRINT = 8.

However, if an option is an array, one can set different values in one group, for example, $PCMCAV RIN(10) = 1.4 RIN(29) = 1.6

$END. RIN (atomic radii for PCM, defined for each atom) has predefined values, which the user may not even know, so it is difficult to set all values in RIN explicitly. In this example, the default values are used for all atoms except for the 10th and the 29th (maybe they are heavy metals without a predefined radius).

In this book, the final $END may be omitted; for example, $FMOPRP NPRINT = 9. If in another place it is suggested to use $FMOPRP NGRFMO(1) = 4, then all such options should be combined into a single $FMOPRP and terminated with an $END, $FMOPRP NPRINT = 9 NGRFMO(1) = 4 $END.

Novice users are advised to study the standard input files in tests/ standard, run the calculations, and examine their output files. By reading the comments in the input files and checking which keywords are set for each type of job, and getting proficient in reading GAMESS output, one can get a thorough basic self-training in using GAMESS for various common tasks, which is a prerequisite for the more complex usages in this book.

2.6 FMO Input Files

It is to be hoped that input files for FMO can be made with a GUI, such as Facio, so that the user does not have to make input files by hand, which would be very tedious. However, a GUI knows only about certain kinds of typical calculations, and it is not uncommon for an experienced user to manually modify an automatically generated input. The FMO-related groups are described here in a concise way to give the important points, and the manual can be consulted for more details.

There are electronic supplementary materials for this book, hosted at https://www.worldscientific.com/worldscibooks/10.1142/ 13063#t=suppl, where a collection of input and output files is made available for download (see **Instructions for Accessing Online Supplementary Material**, p. 295). Most of these files are for Section 6.3, but a few input files used in other Sections are also provided.

2.6.1 *Basis Set ($DATA) and Atomic Coordinates ($FMOXYZ)*

In FMO, $DATA defines a basis set library for each chemical element, without atomic coordinates. That is, to compute $(H_2O)_{20}$ in FMO, $DATA may have just one H and one O, but without FMO, $DATA will have 40 hydrogens and 20 oxygens. Ghost atoms cannot be used with FMO. Some basis sets (such as Dunning) require that spherical atomic orbitals (AOs) be used (ISPHER = 1 in $CONTRL).

There are two main ways to define basis sets in FMO: (1) uniform built-in basis sets defined in $BASIS, suitable for a single layer in FMO and (2) arbitrary basis sets in $DATA. In either case, no coordinates are given in $DATA. Each kind of atom is given once. Up to g-functions may be used in basis sets.

For an example of a uniform basis set, see Table 7, where 6-31G* is defined for a water cluster. For QM methods other than DFTB, the element names (O and H) are ignored but indices after them are used (e.g., in "O-1", O is ignored and 1 is used). A chemical element is identified by its charge in the second column (8 and 1). In DFTB, the first column identifies a chemical species, and this name should match parameter definitions, so that chemical element names should be given in column 1 without any indices, e.g., "O".

Table 7.   Example of a simple basis set definition for FMO (6-31G* for all atoms).

␣$BASIS GBASIS = N31 NGAUSS = 6 NDFUNC = 1 $END
␣$DATA
C1[a]
Job name[b]
O-1 8
H-1 1
␣$END

[a]In FMO, no symmetry may be used (C1 means C_1 point group).
[b]A job name is at most 80 characters, possibly containing spaces. A good practice is to indicate the origin of the initial geometry (such as "from structure 1 of 1L2Y PDB" or "optimized at the level of DFT3/3ob, the energy is −123.456"). The molecular geometry is given in $FMOXYZ.

Table 8.   Example of a 2-layer $DATA with 2 basis sets for O in each layer.

Input	Comment[a]
␣$DATA	
C1	point group C_1
2 layers with 2 basis sets in each	brief description of the job
O.1-1 8	O, basis 1, layer 1
N31 6	6-31G
	Empty lines here and below are required.
O.2-1 8	O, basis 2, layer 1
accd	aug-cc-pVDZ
H.1-1 1	H, basis 1, layer 1
N31 6	6-31G
O.1-2 8	O, basis 1, layer 2
N311 6	6-311G
O.2-2 8	O, basis 2, layer 2
acct	aug-cc-pVTZ
H.1-2 1	H, basis 1, layer 2
N311 6	6-311G
␣$END	

[a]The comments should not be placed in the input.

$DATA for a relatively complicated example is shown in Table 8 (in it, $BASIS is not used). In this example, there are two layers (see Section 4.6) and two O basis sets within each layer.

The complete format of the element field in column 1 of a $DATA entry is $A.i\text{-}j$, where A is an element name that is read but ignored; ".i" and "-j" are optional. Either or both may be omitted (then they are 1). Within each layer, i differentiates between multiple basis sets (here, 1 and 2), whereas j defines layers (here, also 1 and 2). An element field such as "H.2-2" may be at most 8 characters long.

For example, in a protein-ligand complex, the binding site may be assigned to layer 2 (here, 6-311G), and the rest to layer 1 (6-31G). However, anionic functional groups in each layer may be described with a diffuse basis (aug-cc-pVDZ and aug-cc-pVTZ). The choice of the basis set for each atom is made in $FMOXYZ. Using diffuse functions (and large basis sets in general, which means triple-ζ with polarization or larger) in FMO is difficult. Novice users should **avoid** them, and for expert users some ways are described in Section 5.2.

In Table 8, two O basis sets per layer are used, one for neutral groups and another for anionic groups. One should not use a two-basis set setup for systems where each element uses a single basis set, for example, when a heavy metal uses an ECP with its basis set, and light elements use a different all-electron basis set. Such cases should be treated as a single-basis set run, by simply defining a basis set for each element in $DATA.

Summarizing, one can define a uniform basis set as GBASIS in $BASIS, in which case $DATA should list simply atoms without the basis set (Table 7). Otherwise, one can define basis sets in $DATA for each chemical element (Table 8). $BASIS provides convenient ways to define Pople basis sets with polarization (*) and diffuse functions (+) such as 6-31G* and 6-311++G**. $DATA has no automatic way to do this; only the main part can be entered symbolically such as 6-31G (N31 6), whereas the polarization/diffuse functions should be given explicitly. Dunning basis sets such as cc-pVnZ can be conveniently defined by their abbreviated symbols (cc$n$) in both cases.

The format of explicit basis sets is described in the manual (docs-input.txt). Practically, basis sets in the right format can be procured from a basis set library.[11] The empty separator line after each element is required when GBASIS is not used. It is also possible to store basis sets in an external file and refer to the data in it (via user-defined basis set names). This is accomplished by choosing EXTFIL = .T. in $BASIS. The file name is defined as EXTBAS in the script gms-files.csh. This file should be readable from all compute nodes (i.e., usually stored in the home directory on a network disk).

Basis sets in GAMESS are of the Gaussian type, except that DFTB uses Slater basis sets that cannot be defined in $DATA. For DFTB

".*i*" and "-*j*" should be omitted, and $DATA should use the element names such as "C" or "H" (C-1 or CARBON are not accepted in DFTB).

Atomic coordinates are normally given in $FMOXYZ. The format for each line in $FMOXYZ is "*A.i Z x y z*" (one atom per line), where *A* is ignored, and ".*i*" has the same meaning as in $DATA (e.g., C.2 and O.2 can be used for a carboxyl group described with a different basis set). By default, $i = 1$. *Z* is an atomic number or a chemical element name; *x*, *y*, and *z* are Cartesian coordinates. As an example, "COO.2 8 1.0 0.0 –1.20" defines an oxygen atom (8) that should use the second basis set (2); COO is ignored by the program, but may be helpful to the user as a way of a visual aid. 1.0 0.0 –1.20 are Cartesian coordinates in Å. It is impossible to use internal coordinates to define an initial geometry in FMO (no *Z*-matrix).

There is an option to move Cartesian coordinates from $FMOXYZ to an external file (MODIO = 512 in $SYSTEM), which is seldom used. GAMESS can read PDB data in $PDB for defining segments (section 3.5.4), but the atomic coordinates in $PDB are **always** ignored.

2.6.2 *Fragmentation and QM Methods ($FMO)*

$FMO is the most important group for FMO. It defines fragmentation and methods (wave functions, layers, approximations, etc.). This group is big and describes many things, so it deserves a structured description.

2.6.2.1 *Assignment of atoms to fragments*

There are three ways of doing the assignment: (a) automatic, (b) index-based, and (c) list-based. For all three of them, one has to define the number of fragments *N* as NFRAG = *N*.

The automatic way can be applied when all fragments have the same number of atoms, and the atoms in $FMOXYZ are ordered (first fragment 1, then 2, etc.). Most commonly, this way is used for homogeneous molecular clusters. To use this method, NACUT is set to the number of atoms per fragment (for water, NACUT = 3).

For the index-based method, INDAT(i) should be set to the fragment ID to which atom i is assigned. For instance, INDAT(1) = 1,1,2,3,1 means that atoms 1, 2 and 5 are in fragment 1, atom 3 is in fragment 2 and atom 4 is in fragment 3.

In the list-based format, INDAT defines a list of indices. To indicate that this is a list, the first element of INDAT(1) should be 0. After listing all atoms in fragment 1, a separator 0 is placed, then atoms in fragment 2 are listed, another 0, etc. Atoms in the list can be given either as single numbers or as ranges. A range is a pair of a positive and a negative number, so that 3 −6 means atoms from 3 to 6 (a space or a comma should be added before each minus). INDAT(1) = 0 1 −2 5 0 3 0 4 0 defines the same fragments as in the index example above. It should be mentioned here that there is a limit to the number of indices. Ranges should be used as much as possible: instead of 1 2 3 4 5 7 0, it is better to specify 1 −5 7 0.

INDATP is used to define segments (Section 3.5.4) in PA and PAVE. The input format of INDATP is the same as INDAT, except that if INDATP(1) is negative, the rest of INDATP is not read and segments are defined in $PDB.

2.6.2.2 *General FMO settings*

NBODY = n defines the order of the MBE n (FMOn) which can be 1, 2, and 3 (NBODY = 0 is a way to do a quick check run). For FMO0 (isolated state), NBODY = 1 and RUNTYP = FMO0 are set.

In FMO1 only monomers are calculated, and the method has a very limited use. FMO1 gradients can be computed only for NFRAG = 1. The main usage of FMO1 is to do a calculation of excited states (Section 4.5) or properties on a grid (Section 4.9). There is a special use of FMO1, to do *ab initio* calculations without fragmentation using the FMO driver, for example, in order to do PA calculations, accomplished with NFRAG = 1.

FMO2 (NBODY = 2) is the most common way of running FMO. It involves calculations of fragments and their pairs. FMO3 (NBODY = 3) has an additional step of calculating triples of fragments.

2.6.2.3 *Charges and multiplicities of fragments*

A charge and a spin multiplicity ($2S + 1$) has to be set for each fragment, where S is the spin quantum number. The multiplicity applies to the electronic state in SCF. For TDDFT (CIS), the multiplicity of the excited state is chosen in $TDDFT ($CIS).

RHF may only use the multiplicity of 1 (a singlet). Only one MCSCF fragment may be assigned a non-singlet multiplicity. Any ROHF/UHF fragment may have a non-singlet multiplicity (multiple ROHF/UHF fragments are supported), except that if excited states are computed, then only one fragment may be non-singlet.

The fragment charges are defined in ICHARG and the multiplicities in MULT. For example, ICHARG(1) = 0,0,1,−1 defines two neutral, one cationic, and one anionic fragments. MULT(1) = 1,1,3,1 defines three singlet fragments and one triplet. The default is to have all neutral singlets.

2.6.2.4 *QM specifications of fragments*

For each fragment, its SCF type is set in SCFFRG, limited to NONE, RHF, ROHF, UHF, and MCSCF. NONE is used to describe a fixed embedding due to a fragment (see Section 5.2). For example, SCFFRG(1) = RHF,ROHF,RHF,RHF,ROHF defines two ROHF fragments and three RHF fragments. It is impossible to mix ROHF and UHF fragments; MCSCF can only be mixed with RHF (for instance, SCFFRG(1) = RHF,ROHF,RHF,UHF or SCFFRG(1) = RHF,ROHF,RHF,MCSCF are not allowed).

If QM methods are not specified in $FMO, then the method is chosen by SCFTYP and DFTTYP in $CONTRL.

For DFTB, the SCF type should be RHF. For RDFT and UDFT, SCFTYP is RHF and UHF, respectively, and there is no RODFT. For example, UB3LYP is set up as $CONTRL SCFTYP = UHF DFTTYP = B3LYP $END.

ICHARG and IMUL in $CONTRL are ignored in FMO, but for ROHF/UHF/MCSCF, SCFTYP in $CONTRL should be set to one of these non-RHF types, and for DFT, DFTTYP in $CONTRL should be set.

For MP2, the order of the perturbation theory may be given as MPLEVL = 2 in $CONTRL. There is no MP3, etc.; the only choice is MP2. Variations of MP2 such as RI are set in $MP2. CC is defined with CCTYP in $CONTRL. For configuration interaction, only CITYP = CIS is allowed.

2.6.2.5 *Multiple layers*

Multilayer FMO, described in more detail in Section 4.6, requires specifying the number of layers as NLAYER and an index array LAYER(*I*) assigning to what layer fragment *I* belongs. For example, NLAYER = 2 LAYER(1) = 2,1,2,1,1 assigns fragments 2, 4, and 5 to layer 1, and fragments 1 and 3 to layer 2. The QM complexity of layers increases in the ascending order of layers. DFTB may not be used in a multilayer calculation.

For each layer, its SCF type can be set as SCFTYP in $FMO, MP order as MPLEVL, CC type as CCTYP, and DFT functional as DFTTYP. For example, to use RHF in layer 1 and MP2 in layer 2, MPLEVL(1) = 0,2 is used. SCFTYP(1) = RHF,MCSCF is set to define RHF and MCSCF for layers 1 and 2, respectively. TDDFT type for each layer can be specified as TDDFT. Basis sets for different layers are defined in $DATA (see Section 2.6.1).

Multilayer calculations have no exact gradient (except when FD is used; see Section 4.6), and there are various other limitations.

2.6.2.6 *Options for individual fragments and segments*

Packed options for fragments are specified as NOPFRG, described in Table 9. For example, NOPFRG(3) = 129,132 defines the suboptions 128 + 1 for fragment 3, and 128 + 4 for fragment 4; other fragments use the default values (that is, 0).

Suboptions for segments are specified in NOPSEG(*i*): 1 can be set to split the side chain from the backbone of residue *i* (so that two segments are produced for this residue; the backbone remains

Table 9. Fragment-specific suboptions NOPFRG (more values can be found in docs-input.txt).

Value	Meaning
1	Increase the print-out level (useful for plotting MOs in MCSCF and TDDFT fragments).[a]
4	Generate cube file.
64	Use frozen atomic charges for the embedding due to this fragment (see Section 5.2).
128	Apply options 1 and 4 only for the last iteration in the monomer loop.
256	Print R_{IJ} distances (both unitless and in Å) for all $J < I$ if NOPFRG(I) has 256 added to it.
1024	Calculate electrostatic potential on a grid for the transition density in TDDFT.

[a]The idea is to use NPRINT = –5 in $CONTRL, and increase the print-out level to NPRINT=7.

numbered "i", whereas the side chain is appended at the end of the segment list). Such a splitting requires a $PDB.

Names of fragments can be specified in FRGNAM, subject to the usual restrictions (1–8 characters). Names of segments cannot be explicitly defined in the input. When $PDB is provided, segment names are taken from it, by combining two fields, residue name and ID. When side chains are split, the parent name is copied with a prefix "s", whereas backbones keep the parent name.

2.6.2.7 *Subsystem definition*

Fragments can be grouped into subsystems for a more structured presentation of results. There are two usages of this functionality: to skip unwanted dimers, or to do subsystem analysis.

Sometimes the user may like to save time by not computing some dimers (trimers). A typical example is a protein-ligand complex. The user may be interested only in fragment pairs of the type residue-ligand, but have no interest in residue-residue pairs. Or it may be the opposite: there may be several ligand molecules in a protein and the goal is to compute only ligand-ligand interactions in the protein embedding.

For the first kind of pair pruning (to select protein-ligand interactions only), MODMOL = 1 MOLFRG(1) = *I,J,...* can be used, where *I, J*, etc., are the fragment IDs comprising the ligand. Then dimers are limited to those pairs, where one fragment is in the MOLFRG list and another outside the list.

For the second kind (e.g., ligand-ligand interactions only), the user can set MODMOL = 3 and MOLFRG(1) = *I,J,K,...* Then only pairs within MOLFRG are computed (*IJ, IK, JK*, ...).

In the fragment-based subsystem analysis, described in more detail in Section 3.5.8, fragments are assigned to subsystems. For subsystem analysis, MODMOL is set to 8 and MOLFRG(*I*) defines the subsystem to which fragment *I* belongs. For example, in MODMOL = 8 MOLFRG(1) = 1,1,1,1,1,2, the initial five fragments are in subsystem 1, and the last one is in subsystem 2. It can be useful to assign all fragments to subsystem 1 (for example, to define partial energies of residues in an isolated protein following Eq. (17)).

A segment-based SA may also be performed, but in a less detailed way than its fragment analogue. It is accomplished by setting the second vector in NOPSEG. As described above, the first vector in NOPSEG is used for defining segment-specific options; the second vector is used for subsystems. An example: for a polypeptide composed of 5 residues, the last residue is split into backbone and side chain, so there are 6 segments. To define 3 subsystems, NOPSEG(1) = 0,0,0,0,1,0, 1,1,2,1,3,3 can be used. Here, the first vector 0,0,0,0,1,0, defining segment options, means that for residue 5 in the PDB its side chain is split (producing 6 segments total). The second vector of 1,1,2,1,3,3 defines subsystems: segments 1,2,4 are in subsystem 1, segment 3 is in subsystem 2, and segments 5 and 6 are in subsystem 3.

2.6.2.8 *Approximations*

To get good computational performance and reduce the scaling, there are several approximations, summarized in Table 10. All these keywords begin with an "R", which signifies that they refer to distances ("radii"). They function like this: if R_{IJ} > Rkeyword, then the

Table 10.   Keywords related to approximations.

Keyword	FMO*n*	Object	Default[a]	Meaning
RESPPC	*n* > 0	ESP	2.0	Embedding (ESP) with point charges (PC).
RESPAP	*n* > 0	ESP	0.0	Embedding (ESP) with atomic populations (AP).
RESDIM	*n* > 1	dimers	2.0	Electrostatic (ES) approximation for a dimer (DIM) instead of SCF.
RCORSD	*n* > 1	dimers	2.0	HF (DFT) dimer instead of MP2 (TDDFT).[b]
RITRIM	*n* > 2	trimers	1.25,–1,2,2	Skip separated trimers.
MODESP	*n* > 1	ESP	0	Blockwise application of ESP approximations.

[a]Default values depend on input options; here the most common case is given.
[b]The idea is that far separated dimers make no contribution to electron correlation or excited states.

approximation denoted by "keyword" is applied to I,J dimer, where R_{IJ} is the interfragment distance. If a threshold value is zero, the corresponding approximation is not used. When a keyword is negative (e.g., –1), the approximation is applied to all pairs.

Without periodic boundary conditions (PBC), the distance between fragments I and J is computed as

$$R_{IJ} = \min_{A \in I, B \in J} \left(\frac{R_{AB}}{W_A + W_B} \right) \tag{21}$$

where A is an atom in I, B is an atom in J, R_{AB} is the geometric (Euclidean) distance between A and B, W_A and W_B are van der Waals radii of A and B, respectively. The radii can be defined as VDWRAD in \$FMOPRP for each chemical element. As should be clear from Eq. (21), distances in FMO, as well as thresholds to which they are compared, are unitless. For PBC, the expression in Eq. (21) has to be modified. Atom B can be in any neighboring cell, which introduces a loop over cells in computing R_{IJ}.

The reason for dividing by a sum of van der Waals radii is to add a physical perception to geometrical distances. Consider two hydrogen atoms at 2 Å, and two O atoms at the same distance. O atoms may interact stronger. The radius of H and O is 1.2 and 1.4 Å, respectively. Plugging these values, one obtains the distances of 2/2.4 = 0.83 and 2/2.8 = 0.71 for H and O atoms, respectively; the O-O pair is relatively closer than H-H. A relative distance R_{IJ} of 0.5 corresponds to a covalent bond; a hydrogen bond is about 0.8. If there is a detached bond between I and J, one atom (BDA) is present in both fragments, and R_{IJ} is 0.

RESPPC and RESPAP affect the embedding potential (ESP, see Section 5.2). The embedding of fragment I is computed as a sum of contributions from fragments J other than I. If R_{IJ} is larger than RESPPC, then a point charge form of the ESP is used. In DFTB, a point charge ESP is always employed, so RESPPC is ignored for DFTB, although it should still be non-zero. RESPAP is a more elaborate (diagonal two-electron) model of embedding than RESPPC, so a RESPAP value should be less than RESPPC (Figure 8). However, due to its poor performance, RESPAP is almost never used.

RESPPC is typically applied with a sufficiently large value (2.0 or larger), so its influence is relatively weak, but not negligible.

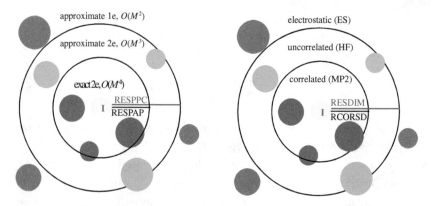

Figure 8. Schematic representation of approximations: ESP (left), and dimers (right). Fragments are shown as filled circles. The blue fragments near fragment I are treated at the best level (exact 2e for ESP, MP2 for dimers). The midway green fragments are treated with a middle level (approximate 2e for ESP, HF for dimers). The farthest red fragments are treated in the most approximate way (approximate 1e for ESP, electrostatic for dimers). The $O(M^k)$ scaling shows the cost in terms of the number of AOs per fragment (M).

RESPPC = −1 defines a point charge embedding in FMO, which is used for treating diffuse functions in molecular clusters. Any non-zero value of RESPPC introduces a numerical error in the analytic gradient, except for DFTB. The RESPPC-related error increases as the threshold value decreases. When the gradient accuracy is of special concern, as in MD, RESPPC can be increased to 2.5 or 3.0. Due to the *n*-mer ESP imbalance problem (Section 5.2), the value of RESPPC is often increased to 2.5 in FMO3. MODESP is an important option for ESP in FMO, described elsewhere (Section 5.2).

The other three thresholds in Table 10 refer to approximations, with exact analytic gradients, and these approximations have a much milder effect on the results than RESPPC. The RESDIM approximation is based on the locality of CT (and exchange), so that for far separated fragments only the electrostatic interaction is computed (and UFF dispersion is added in DFTB).

The RCORSD approximation is based on the locality of the electron correlation, which may have a different distance behavior than the CT (affected by RESDIM). In practice, however, RCORSD in FMO2 typically uses the default value of 2.0, i.e., RCORSD = RESDIM. In MP2/CC, all dimers with the separation longer than RESDIM are computed with the ES approximation; those between RCORSD and RESDIM are computed with HF, and pairs closer than RCORSD are computed with a correlated method (Figure 8).

RITRIM is a vector consisting of four elements, used only in FMO3. Differently from dimers, far separated trimers are not computed at all. The first three elements of RITRIM are used to neglect trimers with a small three-body CT, and the fourth is used for electron correlation. Here, the common usage of the RITRIM algorithm is described (Figure 9), in which RITRIM(2) is not used (set to −1). In a trimer *IJK*, there are three dimers *IJ*, *IK*, and *JK*. The shortest (min for minimum) separation among the three is denoted by R_{IJK}^{min}, and the next in magnitude is R_{IJK}^{med} (med for medium). If R_{IJK}^{min} is larger than RITRIM(1) or R_{IJK}^{med} is larger than RITRIM(3), the trimer is not computed. Because the number of trimers rapidly increases with the values of RITRIM, these thresholds have a substantial impact on the speed (and accuracy) of FMO3.

$$R_{IJK}^{min} \left| \begin{array}{c} J \\ \\ \underline{\phantom{R_{IJK}^{med}}} \\ R_{IJK}^{med} \end{array} \right. K$$

Figure 9. Trimer *IJK* calculation is not performed if R_{IJK}^{min} (minimum interfragment separation among three pairs) is greater than RITRIM(1) or R_{IJK}^{med} (medium separation) is greater than RITRIM(3).

Table 11.   FMO3 levels of accuracy.[a]

Level	RESDIM	RCORSD	RITRIM(1)	Usage
low	2.5	2.5	0.001,−1,1.25,1.25	Nanomaterials.
medium	3.25	3.25	1.25,−1,2.0,2.0	Short-range CT in hydrogen bonds.
high	4.0	4.0	2,2,2,2	Medium-range CT.

[a]RESPPC ≥ 2.5 is usually set for FMO3.

The fourth value, RITRIM(4), is used for correlated methods such as MP2 and CC. If R_{IJK}^{med} is larger than the threshold, the correlated trimer calculation is not performed (but SCF for the trimer may still be done as controlled by RITRIM(1)).

It is thought in general that the values of RESDIM and RCORSD should be increased in FMO3. If this is not done, a situation can occur where CT is ignored in a dimer but computed in a trimer with a comparable separation, which should best be avoided although technically it is permitted. The recommended value of RESDIM in FMO3 is RITRIM(1) + RITRIM(3) or larger, and of RCORSD is RITRIM(1) + RITRIM(4) or larger.

As a simple guide, there are three levels of FMO3 accuracy: low, medium, and high (Table 11). The low level adds three-body effects due to detached bonds, which can be useful in inorganic materials with a high density of such bonds. The medium level also adds three-body effects in regular hydrogen bonds. The high level adds longer-ranged effects related to CT, which may be small per trimer, but accumulate to a substantial sum, because there are many of them. The medium level is most commonly used.

If no approximations are used (i.e., all thresholds are set to 0), then interfragment separations are not computed, which may have various unexpected and possibly disastrous consequences. If the user prefers to compute the distances but avoid approximations, it may be the simplest to set RESDIM to a huge value, such as 1000.

The default values of FMO approximations were chosen for medium basis sets. When larger basis sets are used, with a more diffuse character and longer-ranged QM interactions, larger values of approximation thresholds may be used. One can add 0.5 or 1.0 to the default values, which will also increase the calculation costs. However, because larger basis sets have a worse accuracy without approximations, increasing the thresholds may not improve the overall accuracy.

2.6.2.9 *Analytic gradient*

The exact analytic gradient in FMO requires solving special equations in the self-consistent Z-vector (SCZV) method, which are similar to coupled-perturbed (CP) equations used in the analytic Hessian. These equations are wave function specific, for example, CPHF equations are solved for HF.

By solving these equations, response terms are obtained that arise from the use of fixed monomer embedding in dimer calculations. The extra cost of SCZV is not large, but these equations have to be derived and programmed for each wave function separately, and for each type of the embedding (Section 5.2).

An analytic gradient is needed for geometry optimizations, MD, and semi-numerical Hessian. Whenever available, the exact gradient with SCZV should be used. Otherwise, one has to make do with an approximate analytic gradient.

The details of computing FMO gradients are set in the packed MODGRD option. The recommended options are listed in Table 12. For RMP2, the exact gradient requires RESPPC = 0; an alternative is to use an approximate gradient without this restriction. For analytic gradients, MODESP = 0 in $FMO should be used.

Table 12.   Options for analytic gradients in a single layer FMO.

Method	Options in $FMO[a]	Comments
DFTB	MODGRD = 32	Exact gradient
HF, DFT	MODGRD = 42	Exact gradient
RMP2	MODGRD = 42 RESPPC = 0	Exact gradient
MP2	MODGRD = 10	Approximate gradient
TDDFT	MODGRD = 10	Approximate gradient
MCSCF	MODGRD = 1	More approximate gradient

[a]MODGRD values are for HOP, and 16 has to be added for AFO. SCZV is not available for some combinations of methods, in which case MODGRD = 10 can be used. For DFTB, MODGRD = 32 and 42 give the same results.

One additional factor affects the gradient: the projection. QM calculations in general should be invariant to a rotation or translation of the system (i.e., if 5 Å is added to all atomic coordinates, the results should not change). The translational invariance seems to be always fine, but the rotational invariance is destroyed by using grids, for example, in DFT or in PCM. This produces a numerical noise, and oscillations in geometry optimization. It is common to project out the translational and rotational degrees of freedom by applying a transformation to the gradient. It is done by default for DFT and PCM. The user can change the default choice to the opposite, for example, turn off the projection for PCM by adding 64 to MODGRD.

What happens when a gradient is inaccurate? In a geometry optimization, energy oscillations can occur if the optimization threshold OPTTOL is smaller than the gradient error (neglecting response terms obtained in SCZV incurs a gradient error of about $10^{-4}...10^{-3}$ hartree/ bohr). The energy may go up and down, and the gradient may not decrease for many steps. In MD, the energy conservation can be violated. This, however, applies to NVE, and in NVT the gradient error results in some unphysical force on the system. The error in the gradient accrues in a semi-numerical Hessian, and because the gradient differences are used, the error can be magnified. A visible result of gradient errors is that the translational and rotational frequencies are far from 0.

2.6.2.10 *Array dimensions*

FMO has to guess the size of some arrays and allocate memory before the exact dimension is known. If the default guess is insufficient, which is rarely the case, the user has to adjust these dimensions in the input. By setting the exact dimensions, the memory usage can be reduced. The array dimensions are listed in Table 13. All of these parameters are upper bounds (it is permissible to use a larger value than is actually needed).

MAXKND may have to be increased in the very unlikely case of having 6 different detached bond types (for example, at C, Si, Ge, Sn, Pb, and S atoms) in an AP run (requiring two HMO sets per bond type), resulting in MAXKND = 12. Changing MAXCAO is more common, used for detaching bonds at elements heavier than the first row. For example, for Si-Si bonds, MAXCAO = 9 should be used (a typical Si atom has 9 occupied orbitals, corresponding to 1s, 2s, 2p, 3s, and 3p, where a p-orbital counts as 3). MAXBND has to be increased if there are many detached bonds, which can occur in an inorganic system.

There is no dimension for the total number of atoms in FMO per se; however, non-FMO code used in some RUNTYPs (Table 3) limits the maximum number of atoms to MXATM = 2000. Also, MXATM limits the maximum number of atoms per fragment in any FMO run. There are some other array dimensions. In PCM, MXTS defines the maximum number of tesserae (Section 3.1.3). MEM10 and MEM22 define arrays used to store two files (F10 and F22) in memory (Section 5.8.1).

Table 13.   Options related to array dimensions in $FMO.

Option	Default Value	Meaning
MAXKND	10	Number of HMO sets in $FMOHYB
MAXCAO	5	Number of HMOs per set in $FMOHYB
MAXBND	$2N + 1$[a]	Number of entries in $FMOBND

[a] N is NFRAG in $FMO, the number of fragments.

2.6.3 *Analyses and Properties ($FMOPRP)*

$FMOPRP is the second most important group for FMO after $FMO. It defines options for property calculations, and various additional features. Options used for performing analyses are listed in Table 14. Of these, IPIEDA and MODPAN are most commonly used and the rest are rarely employed.

Options related to calculating MOs and properties on a grid are listed in Table 15. An important packed option in $FMOPRP is MODPRP, whose functionality is described in Table 16.

Table 14.   Options in $FMOPRP pertinent to analyses.

Option	Values	Meaning	Section
IPIEDA	0	No PIEDA (default)	
	1	PIEDA (no polarization)	3.5.5
	2	PIEDA (with polarization)	5.1.1
MODPAN	0	No PA (default)	
	1	PA	3.5.6
EFMO0(1)	[FMO0][a]	Isolated fragment energies for PL0	5.1.1
EPL0DS(1)	[PL0][a]	Polarization energies in PL0	5.1.1
EINT0(1)	[PL0][a]	Interaction energies in PL0	5.1.1
N0BDA	K	Number of detached bonds set up for generating BDA corrections	4.11.3
R0BDA(1)	[FMO][b]	Lengths of detached bonds, used for BDA corrections	4.11.3
E0BDA(1)	[BDA][a]	PIEDA components for BDA corrections for each detached bond	4.11.3

[a]The values should be generated by a preceding run, indicated in square brackets (see the indicated Section for details of FMO0, PL0, and BDA runs).

[b]R0BDA values are written to DAT in most FMO runs.

Table 15.   Options in $FMOPRP related to MOs and grid.

Option	Meaning	Section
MOFOCK	Fock matrix of the whole system (FMO/F).	4.8
NLCMO	Fock matrix of the whole system (FMO/LCMO).	4.8
NGRID	Three numbers of grid points in x, y, and z directions for monomer properties on a grid.	4.9.2
GRDPAD	Extra space around the molecule for designing a grid box.	4.9

Table 16. Property calculations controlled by the packed option MODPRP in $FMOPRP (more values are given in docs-input.txt).

Value	Meaning	Section
1	Save to F10 the electron density matrix in the AO basis for the whole system.	5.6
4	Compute total electron density on a grid.	4.9.1
16	Construct a grid box around the molecular system.	4.9.1
32	Compute molecular ESP (requires suboption 4).[a]	4.9.1
128	Compute spin density for unrestricted methods (requires suboption 4).[a]	4.9.1
512	Use shared memory to store grid arrays instead of replicated memory.	4.9.1
4096	Make gradient of $DFT IDCVER = 3 and 4 exactly analytic by computing the dispersion for the whole system rather than as an MBE.	4.1.2
8192	Add dispersion to ES dimers (not applicable to DFTB).	4.1.2

[a]Add 4 to other suboptions, for example, 32 + 4 = 36.

Table 17. Packed option NGUESS in $FMOPRP (more options are found in docs-input.txt).

Value	Meaning
2	Prepare dimer/trimer initial guess from monomer results (recommended).[a]
8	Save and reuse dimer electronic states in geometry optimizations and MD.[b]
16	Prior to MCSCF dimers, run RHF for dimers and match active orbitals.
64	Dimer/trimer Aufbau principle for selecting occupied orbitals.
128	Do not use restart data of fragments in geometry optimizations and MD.
256	If unconverged, try a different SCF converger with the final orbitals from the unconverged run.
2048	If unconverged, try a different SCF converger with the initial orbitals from the unconverged run (requires 256).

[a]If 2 is not set, the Hückel guess is used.
[b]Implemented for restricted methods.

Several options in $FMOPRP govern convergence, summarized in Table 17. NGUESS = 64 refers to the dimer/trimer Aufbau principle. It means that when the initial orbitals of a dimer (trimer) are constructed, the occupied and virtual MOs of 2 (3) monomers are sorted according to the orbital energy, as a single set, and the lowest orbitals are taken as occupied. When this option is not set, the monomer Aufbau principle is used, which means that in a dimer or a trimer, the MOs occupied in the monomers are also occupied in the initial guess of dimers (trimers). It is rare that this option is used, relevant to the charge instability (Section 4.10.1).

NGUESS = 128 can be set to prevent reusing the converged state of fragments from a previous geometry, pertinent to geometry optimizations and MD. While it is normally a good idea to reuse such data by not adding 128 to NGUESS, in some cases (for AP or very large systems) it is technically impossible to do a restart and then 128 has to be added.

There are various options in $FMOPRP related to convergence, summarized in Table 18. In some cases by modifying a convergence limit, a method selection is done, for example, the distinction between PL vs. PL0 (MAXIT), and PCM[2] vs. PCM[1(2)] (NPCMIT).

The main option governing printing results is NPRINT, described in Table 19. MODPAR defines details of parallelization, listed in

Table 18.   Options in $FMOPRP related to convergence.

Option[a]	Meaning
MAXIT = K	Maximum number of iterations in the monomer loop ($K = 1$ is used for PL0).
CONV = A	Convergence threshold for the monomer loop (default 1e-7).
MCONV	SCF converger in packed format for each FMO step (Table 1).
MCONFG	SCF converger in packed format for each fragment.
COROFF = A	Use HF instead of DFT while the convergence is worse than A (COROFF is analogous to SWOFF in $DFT). COROFF = 0 (disabling the option) should be used for DFT restarts, FMO0, PL0, and in non-FMO runs with NFRAG = 1.
NPCMIT = K	Maximum number of global PCM iterations. $K = 2$ is used for PCM[1(m)].
CNVPCM = A	Convergence threshold A for global PCM iterations in PCM[m], $m > 1$.
NCVSCF = K	Change SCF converger after the initial K iterations in the monomer loop.

[a]Here, K is an integer number, and A is a floating point value.

Table 19.   Packed option NPRINT in $FMOPRP, controlling printing (more choices are in docs-input.txt).

Value	Meaning
0,1,2,3	Level of output (from 0 for the most verbose, to 3 for the shortest).
8	Print atomic charges.[a]
16	Special test run for a thorough check of bond detachment.
32	Increase NPRINT in $CONTRL at the last monomer iteration, for writing out MOs.
64	Print atomic coordinates for each fragment.
128	Do not print individual ES dimer energies.

[a]Atomic charges are not defined for RESPPC = 0.

Table 20.   Packed parallelization option MODPAR in $FMOPRP (more values are described in docs-input.txt).[a]

Value	Meaning
1	Calculate large fragments first.
4	Reduce the number of broadcasts for disk-based F40.[b]
8	Do interfragment distance calculation in the reverse order.[c]
32	Reduce memory requirements in FMO.
64	For restarts (IREST = 2), broadcast F40 by rank 0 to all others.
1024	Use shared memory to store electronic states of fragments.
4096	Accelerate FMO3, which requires that suboption 1 be not set.
8192	Use shared memory to speed up screening in PCM.

[a]The suboptions turned on by default are in bold.
[b]By using a memory buffer, which in some cases can be quite large.
[c]For a better load balancing due to the use of the permutation symmetry in R_{IJ}.

Table 20, and other parallelization options are shown in Table 21 (for their usage, see Section 2.9.3). Miscellaneous other options are in Table 22.

2.6.4 *Covalent Boundaries Between Fragments ($FMOBND)*

$FMOBND defines boundaries by listing pairs of atoms and specifying details of the boundary treatment. In each pair, the first atom is

Table 21.   Parallelization options in $FMOPRP.

Option	Meaning
NGRFMO	Number of groups for each FMO step (Table 1). The first 10 elements are for layer 1, then another 10 elements for layer 2, etc.
MANNOD	Number of nodes in each group for a manual definition of groups, given as a continuous list matching NGRFMO.[a]
LOADBF	Fragments with more AOs than LOADBF are computed with static load balancing on LOADGR groups, set for each FMO step.[a]
LOADGR	Number of groups used in each FMO step for static load balancing of tasks larger than LOADBF.[a]

[a]See an example of usage in Section 5.10.4.2.

Table 22.   Miscellaneous other options in $FMOPRP.

Option	Meaning
PRTDST	An array of 4 elements: 3 unitless thresholds for checking interfragment distances and an energy threshold in kcal/mol.[a] A list of fragment pairs separated by less than PRTDST(1) is printed. A warning is printed for fragment pairs separated by PRTDST(2) or less. Stop if any pair of atoms A,B closer than PRTDST(3) is not defined in $FMOBND as a detached bond and A belongs to a different fragment than B. In the list of pair interactions, print only values larger than PRTDST(4).
IREST	FMO step to be restarted; mainly, 0 (no restart) and 2 (monomer restart).
IJVEC	An array of I,J,K,L,M_{IJK} quintuplets, to read M_{IJK} MOs for n-mer IJK in layer L as $VECi$ (i is the serial number of the quintuplet in IJVEC).
MODORB	Format for file F40: 0 (density), 1 (MOs), or 3 (MOs + energies).

[a]Setting a value to 0 disables the corresponding feature.

always listed with a minus and the second without it, as in "−1 4", which means that a bond is detached between atoms 1 and 4. The detachment is asymmetric, as denoted by the minus: the first atom with a minus is a bond detached atom (BDA) and the second atom in the pair is a bond attached atom (BAA); see Section 4.11 for a description of boundary models.

For each BDA, the user should provide hybrid orbitals, whereas for BAA they are not needed. In other words, if atom 1 is C, and atom 4 is O, then for "−1 4", the user should prepare hybrid MOs (HMOs) for C, and in the case of "−4 1", for O. The orbitals are placed in $FMOHYB. There are two covalent boundary treatments in FMO, hybrid orbital projection (HOP) and adaptive frozen orbitals (AFO).

For HOP, after the pair of BDA/BAA, a list of names is placed, denoting basis set specific HMOs defined in $FMOHYB. For multiple layers, the list should have one set per layer, starting from layer 1. For AP, the list should have two sets: the smaller basis set first, and then the larger one. In all other cases, only one set is specified for the basis set used. In some older input files, an extra MINI set for the initial guess was defined, which is no longer useful. Each HMO set has a label of 1–8 characters.

For AFO, after the pair of atoms, the charge for the model system built around the detached bond can be specified following NONE (denoting a dummy name). For neutral systems, charges and NONE may be omitted. Two examples are given in Table 23.

Table 23.   Examples of $FMOBND.

HOP for AP with 6-31G* and 6-31+G* basis sets	
⌴$FMOBND	
−3 1 C-631*G C-631+G*	Atom 3 (carbon) is BDA.
−56 8 Si-631*G Si-631+*	Atom 56 (silicon) is BDA.[a]
⌴$END	

AFO	
⌴$FMOBND	
−8 4 NONE −1	The model for the bond is an anion.
−35 12	The model for the bond is neutral (no charge is
⌴$END	specified).

[a]Because of the limit of 8 characters, the label for Si set cannot afford to keep G.

2.6.5 *Hybrid Orbitals for Covalent Boundaries* (*$FMOHYB*)

$FMOHYB defines hybrid orbitals for HOP. For multiple basis sets (multilayer and AP runs), one HMO set should be defined per basis set. Each HMO set should have a label, 1–8 character long, matching $FMOBND. There are no default orbitals, and any HOP run with detached bonds should have $FMOHYB defined. See Section 4.11.1 for instructions on how to obtain HMOs.

For each set defined in $FMOHYB, first a label is given, followed by the number of atomic orbitals for the BDA, then the number of hybrid orbitals. After that, HMOs are defined, with two integer indices, 0 or 1. Two combinations are useful: "1 0" and "0 1". "1 0" denotes the orbital of the detached bond, and "0 1" is used for all other occupied orbitals in the BDA. For example, for a carbon atom in the sp^3 hybridization, there should be a single sp^3 HMO (1 0) that describes the BDA-BAA bond; three sp^3 orbitals (0 1) that describe other bonds of BDA, and 1s orbital of BDA (absent for core potentials and DFTB), making the total of 5 (or 4) HMOs.

The order of HMOs in each list is irrelevant. Because of the general limit of 80 characters per line, the list of LCAO coefficients for each HMO may be split into multiple lines. An example is given in Table 24.

Table 24.   An example of $FMOHYB.

Input[a]	Comments
⌴$FMOHYB	
C-6-31G* 3 2	2 HMOs for carbon (3 AOs per atom).
1 0 1.0 0.0 0.0	Detached HMO.
0 1 0.0 1.0 0.0	Other HMO.
Si-6-31G* 8 2	2 HMOs for silicon (8 AOs per atom).
1 0 1.0 0.0 0.0 0.0	Detached HMO.
0.0 0.0 0.0 0.0	Coefficients are split into 2 lines.
0 1 0.0 1.0 0.0 0.0	Other HMO.
0.0 0.0 0.0 0.0	
⌴$END	

[a]This group has an unrealistically small number of orbitals, used to demonstrate the format.

2.6.6 *Model Systems in AFO ($AFOMOD)*

$AFOMOD is used for AFO, when some automatically constructed model systems have to be manually redefined, overwriting the default models. It is necessary to do this when the automatic algorithm generates an inappropriate model, which can happen when bonds on the model boundary (as determined by RAFO) cannot be replaced by bonds to hydrogens (see Section 4.11.2).

$AFOMOD can include multiple models. For each model system, first a number is given, equal to the serial number of the bond entry in $FMOBND. For example, if the user wants to overwrite the model system for the sixth detached bond in $FMOBND, by providing a manually chosen molecular model composed of 4 atoms, then the two leading numbers are 6 and 4. After that 5 numbers are given for each atom: A, Q, x, y, and z. A is the serial number of an atom in the full system, Q is the nuclear charge and x, y, and z are Cartesian coordinates (in Å).

The first atom in each model system should be BDA and the second atom is BAA. Additional atoms, not found in the original system, may be added. For them, the index A should be set to any atom in the original system with a different nuclear charge. For example, to add an oxygen atom not found in the full system, the user should find a non-oxygen atom, which may be carbon number 10, and then set A to 10. An example of a complete group is given in Table 25.

Table 25. An example of a manual definition of a model system H_2CO for AFO.[a]

$AFOMOD				
6 4				
12	8.0	1.0	2.0	3.6
6	6.0	−1.0	2.0	3.2
18	1.0	1.0	−2.0	3.1
5	1.0	1.5	3.2	2.1
$END				

[a]With arbitrary unphysical coordinates used to explain the format. If multiple model systems are needed, all of them should be put into one $AFOMOD, one after another.

2.7 FMO Result Files (LOG, DAT, TRJ, RST, and F40)

Each GDDI group produces an output file (Table 4). By default, the GDDI group with the lowest rank 0 writes its output file to the LOG file, where rungms also writes to. All other groups write their output to files called $job.F06.001, etc., where the last extension denotes the rank (the ordinal number of the compute process). Normally, the main output (that is, the log file) has all the important results. However, if there was a problem, such as SCF of a fragment did not converge, or there was not enough memory, an error message is written in the output of the group that was doing the calculation.

Likewise, a DAT file is created by every GDDI group. The main DAT file, called $job.dat, can contain the following useful data: MOs, grid data in the cube format, and Hessian. DAT files of other ranks are seldom of any use. To save output and DAT files from all groups, some editing of rungms is needed.

Running an MD calculation generates two additional files: an MD restart file (RST) and a trajectory file (TRJ). The restart file is overwritten once per KEVERY steps, so the file size stays small. This file containing atomic coordinates and velocities can be used to restart an MD, for example, to continue the trajectory where it stopped. To use this restart file, some manual editing is needed as described in Section 5.5.2.

The trajectory file contains structures written once per KEVERY steps in MD. The file size grows quickly and can become huge. There are two formats for the trajectory file: native GAMESS format, that only MacMolPlt can read, and multiple point XYZ file, that can be read by many programs such as VMD.[12] The option TRJFMT in $MD determines which of the two is used. It is recommended to use the XYZ format (e.g., TRJFMT = F14o6).

The electronic states of fragments are stored in binary file $job. F40.000, with a copy of this file for each process (MODPAR = 1024 in $FMOPRP replaces this file with a shared memory array). These files for different ranks are equivalent, and any can be used interchangeably. F40 contains either electron density matrices, or molecular orbitals, or both, and optionally orbital energies. This file is used to do PIEDA/PL0 calculations of polarization energies (Section 5.1). MCSCF orbitals can be manipulated via this file.

The format of F40 depends on MODORB and SCFTYP. When these are the same, F40 may be taken from one run and passed to another. MODORB in many tasks should be set to 3, and for simplicity MODORB = 3 can always be set. In this case, only SCFTYP can limit reusing the file. For example, computing a system with RHF and reusing F40 for RDFT is fine, but taking an RHF file and using it in UHF or MCSCF is not. ROHF is compatible with RHF.

FMO0 runs generate $job.F30.000 and $job.F40.000. The latter is overwritten for each fragment, and the former (F30) contains the electronic state for all fragments. For computing polarization energies (Section 5.1), F30 should be saved from an FMO0 run, renamed as F40 and used for restarts. F30 and F40 have the same format.

In the rare usage described in Section 5.6, dictionary file F10 can be saved from an FMO calculation, and reused in a consequent non-FMO calculation.

For MODIO = 512, the initial coordinates in a geometry optimization with RUNTYP = OPTFMO are read from file $job.xyz. and the structure is updated as the optimization proceeds.

Because rungms deletes F10, F30, and F40 files at the end of a calculation, the task of saving and copying them is left with the user, who can modify rungms for the purpose.

2.8 How to Begin Your Own FMO Calculations

Before making your own input files, it is useful to study some samples of input files provided in this book (Section 6.3). More examples are found in tools/fmo/samples (note that they span a long development and may not represent the best choice of options for the latest version).

For making FMO input files, the most advanced software is Facio,[13] which supports both automatic and manual fragmentations, a variety of RUNTYPs, and many kinds of FMO calculations. An alternative tool is FU,[14] which has an advanced support of biochemical functionality, but a limited engine for making FMO input files.

Visualization of molecular orbitals of fragments can be conveniently done using MacMolPlt,[15] which requires that appropriate options be set in FMO in order to write out relevant information (Section 4.8). The user can manually excise output for fragments of interest from an FMO output file and save fragment portions to separate files, one fragment per file. These files can then be read into MacMolPlt.

To process MD trajectories, VMD[12] is convenient. It can read a trajectory file generated with TRJFMT in $MD (as a multiple structure XYZ file). Some initial preparation of data for PBC calculations can be performed with Mercury,[16] such as replicating symmetry-related cells.

An initial structure preparation for QM calculations is a very important step. A force field may be quite tolerant to having too long or too short bond lengths in the initial structure, and the strain can be relieved after a few optimization steps with each bond represented by a predefined spring. On the contrary, for a QM calculation an elongated distance can lead to a diradical character, and calculating it with RHF or RDFT can result in a divergence fiasco. Needless to say, a QM structure should have all necessary hydrogen atoms, so that for many proteins from PDB database it is necessary to add missing hydrogens, which can be done with Facio or other tools such as PDB2PQR.[17] A force field optimization can be used for refining the structure. Some QM methods (such as DFTB) are more forgiving of faults in the structure and can be used even on a relatively coarse geometry.

It is best to avoid counterions unless they are absolutely necessary, and in this case there are some points of concern (Section 4.11.5). Various interaction analyses are available only for continuum solvent.

When all atoms are in place, an FMO input file for GAMESS can be made using Facio[13] or FU.[14] FMO results can be visualized by reading GAMESS output into these programs, or into PyMOL[18] plugin pyProGA.[19] The latter program can visualize PIEs and protein residue networks, which provide a composite way of analyzing protein-protein complexes based on both topological connectivity and interactions.

2.9 Parallelization Strategy

2.9.1 *The Golden Rule*

The technical aspects of running FMO with rungms are described in Section 2.4. Now the question of choosing the best strategy is addressed. For DFTB, the general mode of running is 1 core per group, so there is no question of choosing the group size, it is always 1. But for other methods, choosing a group division is an important issue.

The golden rule of how many GDDI groups to define for parallel efficiency is quite simple: use $M/3$ groups, where M is the number of tasks. The number of tasks at each step in FMO differs. There are N fragments, so that $M = N$ for the monomer step **1** (Figure 2). In the commonly used RESDIM approximation (Section 2.6.2.8), most dimers are computed with a fast Coulomb model (far separated dimers) and the rest with SCF. The number SCF dimers and trimers should be taken as M for steps **2** and **3**, respectively.

Figure 8 shows that for each monomer, the number of SCF dimers (limited by the radius RESDIM) is independent of N, so that the number of such dimers per monomer is a constant (a_2). Likewise, the number of trimers per monomer is independent of N, with a larger constant a_3 than for dimers. The number of SCF monomers, dimers, and trimers is thus N, $a_2 N$, and $a_3 N$, respectively.

The golden rule is to use $N/3$, $a_2 N/3$, and $a_3 N/3$ groups for monomer, dimers, and trimers, respectively, which is accomplished with assigning these numbers to NGRFMO(1). The values of prefactors a_2 and a_3 depend on the geometry and thresholds (RESDIM and RITRIM). Smaller fragments tend to have larger prefactors (as in water clusters), whereas some directionality in geometry reduces the prefactors (if all fragments form a chain or are in a flat arrangement). The numbers of SCF dimers and trimers are printed in an FMO run, and they can be used for designing an efficient parallelization.

What are the systematic general exceptions to the golden rule? If there is a large variation in the cost of tasks (some tasks are fast, some are slow), then a better solution may be found rather than that given by the golden rule (see some examples in Section 5.10.4). The golden rule should not be applied to DFTB because for that method the number of groups should be equal to the number of cores (1 core per group), the same number of groups for monomers, dimers, and trimers.

The number of groups cannot exceed the number of nodes. When the numbers of nodes and fragments are small, then it may be beneficial to use 1 group for monomers, and not $N/3$, to minimize synchronization loss.

Each layer can have a different number of fragments and NGRFMO should be defined for each layer separately. By design, in layer 1 all N fragments are always calculated. In layer 2, only fragments assigned to this layer are recomputed. Although only 3 values are used in NGRFMO above, there are in fact 10 elements of NGRFMO per layer, of which only 7 are used. NGRFMO(1:10) is for layer 1, NGRFMO(11:20) for layer 2, etc.

2.9.2 *Uneven Task Sizes*

In FMO-MCSCF, only one fragment is MCSCF and the rest is RHF. The calculation cost of an MCSCF fragment may be many-fold that of all other RHF fragments, so that for a better efficiency, two group sizes can be defined for monomers in FMO-MCSCF: NGRFMO(1) is used for the only MCSCF monomer, so that NGRFMO(1) = 1, whereas NGRFMO(6) is used for RHF monomers. Likewise, for dimers in FMO-MCSCF, one can specify NGRFMO(2) for MCSCF dimers, and NGRFMO(7) for RHF dimers.

In TDDFT, there is one TDDFT monomer (except for FRET), but there may be multiple TDDFT dimers (including the TDDFT fragment). There is no separate setting for DFT and TDDFT monomers, so that NGRFMO(1) is used for all of them. For FRET there may be many TDDFT fragments and NGFRMO(1) can be set to their number. CIS at present is only implemented at the FMO1 level, but in terms of choosing the number of groups it is similar to TDDFT.

For ROHF and UHF (UDFT), there may be one or several open-shell fragments. There are no special group settings for them. Open-shell fragments typically are more difficult to converge than closed-shell ones, and also the cost of open-shell methods is roughly twice larger for the same size.

The cost of doing MP2 for a fragment scales more steeply than the cost of RHF, and if there is a large variety in the fragment sizes, choosing the group size for MP2 may require more thought, with the likely decision to use fewer groups than for RHF.

A somewhat challenging case is when there is one very large fragment (such as a ligand treated as 1 fragment) and many small ones (water or amino acid residues). In this case, semi-dynamic load balancing can be used for better parallel efficiency (*vide infra*).

2.9.3 *Load Balancing*

Load balancing refers to the way of distributing tasks among "workers". In FMO with GDDI, there are two kinds of workers, groups (upper level) and cores in each group (lower level). Consequently, there are two different levels of load balancing: among groups, fragment calculations are distributed, whereas among cores, work in a fragment is divided, for example, integrals.

For the lower level (intragroup), there are two types of load balancing strategies, static and dynamic, specified as SLB and DLB in BALTYP of $SYSTEM, respectively. In the static load balancing, the algorithm of distributing work is predetermined and does not depend on the actual progress. Typically, the index of the work batch is used for dividing the work. For example, if there are 6 tasks and 3 workers, worker 1 will do 1st and 4th, worker 2 will do 2nd and 5th and worker 3 will do 3rd and 6th tasks, irrespective of task sizes and the speed of doing them. This type of load balancing may work well if tasks take the same time. As it is never the case, this static load balancing is seldom used because its alternative is more efficient.

The other choice is dynamic load balancing. In it, master data server is assigned the task of keeping track of work assignment. When any compute process is free, it asks for an index of the next task. This strategy is usually better than the static one, because it can handle uneven task sizes (while one worker is doing a large task, another can finish several small ones), and if nodes are assigned partwise on a public cluster, their performance may be affected by other jobs. Nodes may have different CPUs or memory type, so that one node can do more work than another (as in grid computing).

For the higher level (intergroup), there are three choices of load balancing in $GDDI, static (BALTYP = SLB), dynamic (BALTYP = DLB), and semi-dynamic (also BALTYP = DLB). The distinction between the latter two is done via other options. Static and dynamic choices are just like for the intragroup load balancing: in the static approach, fragments are divided among groups according to their index, and in the dynamic case, when a group finishes one fragment, it asks for another one.

The third type, semi-dynamic, works as follows. All tasks larger than a threshold are done statically on a predefined set of large groups (for this load balancing, a manual selection of group sizes is usually done in MANNOD of $FMOPRP). When all static tasks are done by large groups, these groups do the rest of tasks with dynamic load balancing. Small groups use dynamic load balancing.

This mode was designed to deal with the difficult case of a few large fragments and many small ones. Imagine having 1 elephant and 99 bugs (Figure 10). If there are 10 ants to carry them, and one task is assigned per ant, the ant carrying the elephant will take forever. To deal with the elephant, the following strategy is used. First, ants are divided into uneven groups, one team of 9 ants (i.e., a GDDI group with 9 nodes) and another team of 1 ant (i.e., 1 node).

The threshold LOADBF defines the size of work considered large, and it is set to be slightly less than the elephant weight (for fragments, size is measured in atomic orbitals). LOADGR defines the number of large groups assigned to do large work (carry the elephant). Note that if the static load balancing were not used for large tasks, then there would be no guarantee that large teams get large tasks, because for purely dynamic load balancing, whichever team asks first gets the next task. The smaller team does small tasks, and carries the bugs one by one. If the team of 9 ants manages to carry the elephant home and present it to the queen while there are still bugs to carry, the larger team will also carry bugs (one at a time) along with the smaller team.

The reader may come up with a better strategy of dealing with the elephant, such as let 10 ants carry the elephant first and then split into 10 teams and carry bugs (this is what NGRFMO(1:2) and NGRFMO(6:7) do for MCSCF). An alternative plan of forming just 1

tasks **worker groups**

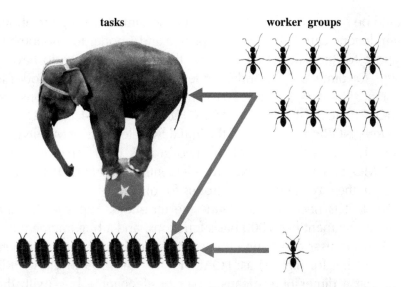

Figure 10. Example of semi-dynamic load balancing. 10 workers are divided into NGRFMO(1) = 2 groups, with 9 and 1 workers (MANNOD(1) = 9,1). If LOADGR(1) = 1 is set, then one large group (of 9 workers) does tasks larger than LOADBF(1), set to be slightly below the size of the large task (999 kg). Large groups (here, only 1 such group) do large tasks with static load balancing (blue arrow), and after that, they do small tasks with dynamic load balancing (red arrow). Small groups do small tasks (one by one) dynamically. The reader is spared the sight of 99 bugs mentioned in the text and only the first 10 are shown (possibly *Armadillidium vulgare*). Tasks are not shown to scale.

team with 10 ants and sticking with it for all tasks may work well for the elephant, but when 10 ants try to carry a small bug, they may get in the way of each other and waste much of their formic energy. Small tasks are best done by small teams.

The semi-dynamic plan can also be used when one fragment is computed with a very expensive method such as TDDFT or MCSCF, whereas other fragments are computed with DFT or RHF (because the cost is determined not only by the size but also by the QM method).

How is the semi-dynamic load balancing accomplished in terms of input options? First, it is necessary to do a manual group definition. This is done with MANNOD in $FMOPRP. Every step in FMO has its own number of tasks and its own set of groups to do them. In FMO2, there are two main sets of tasks, monomers and dimers. So MANNOD should be set separately for monomers and dimers. Taking the above example, one can define two groups of 9 and 1 nodes for monomers. Suppose for the sake of an example that one would like to have 3 groups for dimers, 4, 4, and 2 nodes (the total number of nodes is constant, 9 + 1 = 4 + 4 + 2). Then the node division options are $FMOPRP NGRFMO(1) = 2,3 MANNOD(1) = 9,1, 4,4,2 $END.

Here, NGRFMO(1) is 2, so the initial two elements of MANNOD, 9 and 1, define the sizes of the two groups used for monomers. NGRFMO(2) is 3, so the next three elements of MANNOD, 4, 4, and 2, define the sizes of the three groups for dimers.

Next, it is necessary to define the threshold. Suppose that the elephant fragment has 1000 basis functions, and a bug fragment has 10, then the user can define the two thresholds (one for monomers and the other for dimers) as: LOADBF(1) = 999,1009 in $FMOPRP. Note that a dimer here means a pair of elephant + bug (with the weight of 1010). The thresholds are set slightly below the actual sizes. Any task that is larger than the threshold is treated with the static load balancing done by the special large teams. To have 1 special team of 9 nodes for monomers and 2 special teams of 4 nodes for dimers, LOADGR(1) = 1,2 is used. These special groups (1 group for monomers and 2 groups for dimers) are always the first groups in the group list, defined in MANNOD, so large groups should be defined first. For MANNOD(1) = 9,1, 4,4,2, the only large group (LOADGR(1) = 1) has 9 workers for monomers, and for dimers, there are 2 large groups (LOADGR(2) = 2) with 4 workers each.

It is usually beneficial to sort tasks according to their size. Unsorted tasks complicate load balancing. Imagine that the tasks of elephant + 99 bugs are not sorted. Maybe there are 35 bugs, then the elephant, and then another 64 bugs. It is difficult to deal with distributing work in this case. It is customary to sort tasks in descending order of size. This is done by setting 1 in MODPAR in $FMOPRP.

Now, selecting the group size to maximize performance is actually an even (much) more complex task than may be supposed by reading the above description. There is also an issue of synchronization loss (Figure 7). At some steps, all groups should meet (a synchronization point). Those that reach it first wait for others. This loss may be very substantial.

CHAPTER 3

Basic FMO Calculations

3.1 Treating Solvent

There are three ways to describe solvent in FMO: (1) as explicit QM fragments, (2) as explicit EFP fragments, and (3) as implicit continuum (PCM or SMD). For interaction analyses it is best to use only implicit continuum water (geometry optimizations may use explicit water more lavishly). In analyses, a few explicit water molecules may be added if they are indispensible, for example, if some water is in the binding pocket between a ligand and a protein (a water bridge).

3.1.1 *Explicit QM Solvent*

Solvent molecules can be computed as FMO fragments. The explicit treatment is potentially the best but there are several problems with it. It may be necessary to locate the global minimum and do a conformational sampling to get the entropy. Another problem is that water has large many-body effects and for a better accuracy FMO3 may have to be used. The third problem is that analyzing solvent effects for explicit solvent may be difficult.

Explicit QM fragments for bulk solvent can be used in MD. An alternative that lessens the cost is to use explicit water with EFP, described next.

3.1.2 *Effective Fragment Potential (EFP)*

EFP water is in some ways a more practical approach than explicit QM water. However, combining EFP with many analyses and options in FMO is not implemented, and the use of EFP is limited to geometry optimizations, MD, a special analysis IEA (Section 3.5.8), and a search of the global minimum of solvent molecules with a Monte Carlo approach, RUNTYP = GLOBOP (the default set of options in $GLOBOP is reasonable).

An example of an EFP section in an input file is shown in Table 26. Only the first generation of EFP, EFP1, may be used. An EFP fragment has an internal predefined rigid geometry, and a fragment can only move as a whole (a user-provided geometry of EFP fragments is internally adjusted). Energy and its analytic gradient can be computed for FMO/EFP.

The most important option for FMO/EFP is NLEVEL in $FMOEFP. It defines whether EFP multipole moments are induced by monomers (NLEVEL = 2) or by monomers, dimers, and trimers, depending on n in FMOn (NLEVEL = 1). The former is faster but the latter is more accurate. NLEVEL = 2 should be used for analytic gradients.

The coordinates and EFPs can be defined as shown in Table 26. EFP may be either a genuine QM-derived EFP1 model, with two built-in choices of H2ORHF and H2ODFT in $EFRAG, or a force field of SPC, SPCE, TIP5P, TIP5PE, or POL5P. For EFP1, atoms have a Z prefix (as ZO1 in Table 26), and for force fields, no prefix (O1, etc.). FMO/EFP includes electrostatic embedding from EFP as a kind of

Table 26.   **Definition of EFPs (two water fragments with built-in potentials).**

Input
⌴$EFRAG
COORD = CART
FRAGNAME = H2ORHF

ZO1	−0.4268510404	−0.9690768853	−4.8743291880
ZH2	0.4877150242	−0.8637769545	−5.0825581296
ZH3	−0.9015130436	−0.8822397708	−5.6855219840

FRAGNAME = H2ORHF

ZO1	−1.7703265745	0.5460380520	−2.9215544710
ZH2	−1.2936075996	0.1670076072	−3.6426324106
ZH3	−1.6375357581	1.4792969212	−2.9692297913
⌴$END			

QM/MM. Non-electrostatic QM-EFP interaction can be computed only for genuine EFP1.

It may be possible to generate EFP1 for molecules other than water. No link atoms may be used for FMO/EFP interface.

3.1.3 *Continuum Solvent Models (PCM and SMD)*

Continuum models of solvation are the most convenient way to include solvent, given their extensive development and support in many analyses. PCM and SMD are very similar in many ways, so they are described together. In both models, the whole QM solute is placed in a cavity (Figure 11). The cavity is constructed as a union of atomic spheres (each atom is assigned a radius for the purpose), with overlapping parts removed. The surface of the cavity is divided into pieces called tesserae. Each tessera has a center, in which a single point charge is placed. The apparent solvent charges (ASCs) on tesserae are determined by the potential from the QM solute, and ASCs polarize the solute via a term added to the Fock matrix.

Although PCM is most often used for representing solvent as continuum, it is also possible to represent other kinds of media, in particular, protein environment in a truncated model, where some protein atoms are treated quantum mechanically and the rest are replaced by a continuum with a suitable dielectric constant such as $\varepsilon = 4$ (defined as EPS in $PCM).

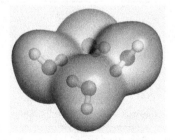

Figure 11. Schematic representation of a solute in a solvent cavity with charges on the surface; the solute and solvent mutually polarize each other to self-consistency. Positive and negative induced solvent charges are shown as red and blue surfaces, respectively.

In FMO, there is a single total cavity for all fragments, and each fragment is polarized by the solvent charges on this total cavity. It is a troubling situation when a solute has remote parts, encapsulated in disjoint subcavities. It is somewhat questionable whether PCM can be used for such a case. Technically, the electrostatic and ICAV = 1 models work, but IDISP = 1 does not, and if multiple subcavities are detected, the IDISP model is turned off by internally setting IDISP = 0 (it does not apply to SMD, where IDISP is never used). It may be unnoticed by the user, because the run continues. Multiple subcavities may indicate that the molecular geometry is not appropriate to begin with.

FMO can be used with two electrostatic models, conductor (C) and integral equation formalism (IEF). SMD is chosen by setting SMD = .T.. A comparison of PCM vs. SMD is given in Table 27.

In PCM, parameters are available only for selected chemical elements, but SMD has a complete coverage. However, SMD requires a large memory quadratic in the number of tesserae vs. small linear arrays in PCM. Some useful options are shown in Table 28.

IEF-PCM has to solve two matrix equations (*vide infra*), and C-PCM only one. C-PCM is a good choice for polar solvents (such as water) with a large dielectric constant ε, whereas IEF-PCM should be used for non-polar solvents or protein medium with a small ε. However, for some rare combinations of options only C-PCM may be used.

Table 27.   Summary of continuum solvent models PCM and SMD.

Property	PCM	SMD
Electrostatic models	C,IEF	C,IEF
Escaped charge ICOMP	0,1,2[a]	0[a]
Number of predefined solvents	17	178
Predefined atomic radius sets	Many[b]	1
Non-electrostatic models	ICAV, IDISP[c]	CDS[d]

[a]0 means that the option is disabled. ICOMP = 1 and 2 have no analytic gradient.
[b]Three sets are built in, and the user can also manually define radii.
[c]For IDISP, parameters are available only for a few elements.
[d]Parameters are predefined for most elements (up to Lr).

Table 28. Useful PCM and SMD options.[a]

Group	SMD	Options	Meaning
$PCM	Yes	IEF = −10	Conductor model (C-PCM),
	Yes	IEF = −3	Integral equation formalism (IEF-PCM).
$PCM	Yes	SOLVNT = WATER[b]	Solvent is water.
$PCM	No	ICOMP = 2	Renormalize solvent charges because of charge escape.
$PCM	No	ICAV = 1 IDISP = 1	Non-electrostatic models.
$PCM	Yes	IFMO = −1	Choose PCM<1>.
$PCM	Yes	MXTS = X^c	Number of tesserae.
$PCMCAV	No	RADII = VANDW	Predefined van der Waals atomic radii (other choices: SUAHF, DFTB33OB).
$PCMCAV	No	RIN(1)	Manually defined atomic radii.
$DISREP	No	DKA(I) = X RWA(I) = Y	IDISP parameters for atom I.

[a]All options apply to PCM; the SMD column indicates if the option also applies to SMD.
[b]Water is chosen as an example. See docs-input.txt for other options.
[c]Maximum number of tesserae is guessed in the program; if the guess is insufficient, the user should increase the value by defining X.

PCM and SMD are based on the assumption that solute is inside the cavity. This assumption is violated by the electron density that decays beyond a finite cavity. The escaped solute charge violates the solvent charge normalization. To deal with it, a charge renormalization can be done with ICOMP = 2, for which there is no analytic gradient. This problem is more severe for diffuse basis sets, whereas for medium basis sets it is minor. A proper charge normalization is important for obtaining accurate local dielectric constants ε_I of charged fragments.

In contrast to SMD, where atomic radii are fixed, the user of PCM has to decide what radii to use, and the radii have a major impact upon results, usually, more than all other options. The main built-in set is that of van der Waals radii (VANDW), defined for most chemical elements. The alternative set of radii is SUAHF, defined only for some common elements. A distinctive feature of the latter is that hydrogen radii are set to 0.01 Å, which reduces the total number of tesserae and permits smoother geometry optimizations and MD simulations. In literature one can find other sets, and atomic radii for them can be defined for each atom in RIN of $PCMCAV.

There is a special set of PCM radii fitted for DFTB3 with 3ob parameters, chosen with RADII = DFTB33OB. To use this set, $PCM ICAV = 0 IDISP = 0 IEF = −10 $END and $PCMCAV ALPHA(1) = 1.0 $END should be used.

If a radius for an atom is not defined, the user can set it. For example, if atom number 34 in $FMOXYZ has no predefined radius, it can be set with RIN(34) = 2.5 as 2.5 Å. Such RIN definitions can be done multiple times for all needed atoms.

In PCM (but not in SMD), the radii thus defined are used with a multiplicative constant ALPHA, equal to 1.2 for electrostatics (it means that in constructing the cavity used for inducing solvent charges, each atomic radius is multiplied by 1.2). The electrostatic, IDISP = 1 and ICAV = 1 terms in PCM require three different cavities using appropriate multiplicative constants. In contrast, SMD has a single cavity for its predefined radii.

Both PCM and SMD have to solve a linear matrix equation to get solvent charges $\mathbf{q}$ (a vector) from the solute potential $\tilde{\mathbf{V}}$ (also a vector),

$$\mathbf{Cq} = -\frac{\varepsilon - 1}{\varepsilon}\tilde{\mathbf{V}} \qquad (22)$$

where $\mathbf{C}$ is a square matrix (C_{ij} are dependent on the distance R_{ij} between two tesserae i and j), and ε is the dielectric constant of the solvent.

Because the linear size of the matrix $\mathbf{C}$ (the number of tesserae) may be large (for a protein with 10,000 atoms it may be ~50,000), the equation is solved iteratively, which can take 50–100 iterations.

In solution, the solute and the solvent polarize each other self-consistently. For two levels, PCM[1] and PCM<1>, this self-consistent loop is integrated into the monomer loop, whereas for higher levels an extra solvent loop is used (Figure 2). PCM<1> is recommended as a rule (PCM[1(2)] may be used for strongly charged

Table 29. Levels of FMO/PCM, in the expected order of increasing the accuracy (everything equally applies to SMD).

Name	Potential $\tilde{V}$	Constant a[a]	IFMO in $PCM
PCM[0][b]	One fragment, iterative	Not used	−10
PCM[1]	Total, one-body, iterative	1	1
PCM<1>	Total, one-body, iterative	2	−1
PCM[1(2)]	Total, two-body, non-iterative[c]	1	2
PCM[2]	Total, two-body, iterative	1	2
PCM[1(3)]	Total, three-body, non-iterative[c]	1	3
PCM[3]	Total, three-body, iterative	1	3

[a]The constant a is applied to multiply $\Delta E_{IJ}^{\text{CT-es}}$; see Eq. (14).
[b]PCM[0] is used for monomer calculations, so constant a is not used.
[c]In PCM[$l(m)$], the l-body iteratively converged potential is used first, followed by a single full iteration of the m-body potential. For these methods, NPCMIT = 2 in $FMOPRP is also added to an input.

systems, such as DNA). A PCM[1(2)] and a PCM[2] calculation is roughly 2 and ~8 more expensive than PCM<1>, respectively. The partial solvent screening model and analytic gradients with SCZV are only available for PCM[1] and PCM<1>.

The difference between levels of PCM in FMO (Table 29) is the order of the MBE of the solute potential $\tilde{V}$ that induces ASCs, and also in the factor a for explicit solvent-related dimer terms.

$$\tilde{V} = \sum_{I=1}^{N} \tilde{V}^{I} + \sum_{I>J}^{N} \Delta\tilde{V}^{IJ} + \sum_{I>J>K}^{N} \Delta\tilde{V}^{IJK} \qquad (23)$$

In PCM<1> or [1], $\tilde{V}$ is obtained by neglecting dimer and trimer terms, which is why PCM can be integrated in the monomer loop (Section 2.1.1).

PCM[l], PCM<l>, and PCM[$l(m)$] calculations can be combined with FMOn for $l \leq m \leq n$. In other words, it is possible to use a lower level in PCM than in FMO, for example, FMO2/PCM<1> is the most commonly used method, in which ASCs are determined from monomer potentials and used in FMO2. In PCM[0], $\tilde{V}^{I}$ is used as $\tilde{V}$ in solving Eq. (22) for each fragment I separately.

Table 30.   Packed MODPAR option in $PCM (more values are in docs-input.txt).

Value	Meaning
1	Sphere-wise, rather than tessera-wise parallelization.[a]
8	Partial screening model (see Section 3.3).
16	Print tessera coordinates and charges.
64	Accelerate the iterative solver in PCM by prestoring data in memory.
256	ASC restart in geometry optimizations and MD.

[a]This refers to how work is divided in parallel, by assigning spheres or tesserae to workers.

MODPAR in $PCM is an important option, described in Table 30. Other options in PCM are very rarely used. MTHALL in $TESCAV chooses the method for dividing an atomic sphere into tesserae. The recommended method is FIXPVA, chosen with MTHALL = 4, which is the default. NTSALL in $TESCAV defines the number of tesserae per atomic sphere. The recommended value is 60, which is the default.

3.2 Desolvation Penalty in Binding (SBA)

In a binding process (see Section 3.5.8), two solvent-related energies are subtracted. Let us consider an isolated ligand that binds a protein. This isolated ligand is solvated from all sides, but in order to bind a protein, some solvent has to go away from the surface of the ligand (and protein), as shown in Figure 12. This is a partial desolvation (a full desolvation is used for defining the solvation energy, and it is not dealt with here).

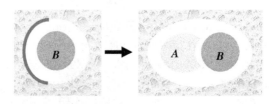

Figure 12. Partial desolvation of *B* in complex formation. The solvent surface lost by *B* in the complex *AB* is shown as the red arc.

There are several approaches to computing the desolvation energy in FMO. The simplest one (the direct approach) is to equate the change in the solvent-related energy ΔE_I^{solv} in the complex formation to the desolvation energy of fragment I, whereas the change in the screening ΔE_{IJ}^{solv} may be either counted also as a desolvation, or, according to Eq. (4), considered a part of the screened solute energy.

There are several more complex models of desolvation for FMO, which are introduced here for completeness. They are indirect approaches, with an intermediate state, in which the binding of A and B in solution is taken to be a multiple step process: a partial desolvation of A, a partial desolvation of B, and the binding of desolvated A and B (the intermediate state is that of isolated A or B computed in the cavity of AB). In the symmetric binding analysis (SBA), the binding energy is computed as a sum of three terms,

$$\Delta E = \Delta E^{desolv,A} + \Delta E^{desolv,B} + \Delta E^{AB} \qquad (24)$$

where the desolvation energy of A ($\Delta E^{desolv,A}$) is computed as the energy of A in the cavity AB minus the energy of A in its own cavity minus the cavitation energy of the empty cavity of B (likewise, $\Delta E^{desolv,B}$ is defined). ΔE^{AB} is computed as the difference of the energy of solvated AB minus the energies of A and B, each computed separately in the total cavity AB, plus the cavitation energies of empty cavities of A and B (to avoid double counting).

There are two further modifications of SBA: asymmetric binding analysis (ABA) and symmetric binding analysis with separated cavitation energy (SBAC). ABA does not separate the desolvation energies of A and B. Moreover, there are fragment-wise formulations of these models, using the many-body energy decomposition.

To enable such calculations, there is a very special (almost never used) way of running FMO/PCM with dummy spheres, that is, with empty atomic spheres (without atoms inside). These spheres have point charges on their surface. The use of dummy spheres is accomplished with defining NESFP and MXSP in $PCM, both set to the total number of spheres (of real and dummy atoms together); that is, NESFP and MSXP should be equal to the number of atoms in the complex *AB*. In addition, XE, YE, ZE, and RIN in $PCMCAV should be set for NESFP atoms (Cartesian coordinates and radii). Several PCM calculations should be performed and the results combined according to the chosen model.

The extra step of running intermediate calculations with empty spheres may unduly complicate the protocol of analyzing the binding. Using the simplest direct desolvation method (the desolvation computed as the sum of the changes in monomer terms ΔE_I^{solv}) may be a good practical approach.

3.3 Solvent Screening

The solvent screening is the extent in which the solvent explicitly affects interactions in the solute (the solvent also implicitly affects the solute via polarization). The most convenient way to add and analyze solvent screening effects is to use a continuum model (PCM or SMD). The main type of screening is electrostatic. However, using the same methodology, one can also define non-electrostatic screening in PCM (this screening is short-ranged and weak). A non-electrostatic screening may sound like an oxymoron; its meaning is a resistance of the solvent with respect to molecular forces in the solute.

The electrostatic screening is related to the dielectric constant ε describing the medium (solvent). There is a global value of ε, corresponding to the bulk solvent. In FMO, local values of ε in the vicinity of each fragment I are also calculated (fragment-specific ε_I values based on the induced solvent charges), which reflect the chemical composition and the actual charge distribution on the solute/solvent interface. Because the screening in FMO takes into account the local electronic structure, it can describe, rich in complexity, details of the solute/solvent interface. There are three types of local screenings, a regular screening for $\varepsilon_I > 1$, a negative screening for $\varepsilon_I < 0$, and the antiscreening for $0 < \varepsilon_I < 1$.

The screening in PCM/SMD (Figure 13) is obtained as crosswise solute-solvent interactions, which can be electrostatic and non-electrostatic. It can be shown that in the idealized case of two ions far separate from each other, the screening model does give the expected effect of

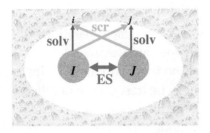

Figure 13. Schematic representation of solvated fragments *I* and *J*, solvent charges *i* induced by *I*, solvent charges *j* induced by *J*, as well as solvent-related energies ΔE_I^{solv} and ΔE_J^{solv} (solv, blue arrows) and solvent screening (scr, ΔE_{IJ}^{solv} green arrows show two contributions to it).

reducing the electrostatic interaction between ions by the global factor of ε. In a realistic case, because of the interference (coupling) between close fragments, the screening (the reduction of the interaction in solution) is a much more complex, local, environment-dependent phenomenon.

A local model and a partial model of screening are available for FMO. They differ in the definition of induced solvent charges (schematically denoted by *i* and *j* in Figure 13).

In the local model, the solvent charges from SCF are used to compute the screening, by dividing the total vector of charges **q** into fragment sections (corresponding to charges in the *local* vicinity of each fragment). This is the simpler model that does not require solving additional PCM equations, so it is the faster model of the two. This model works very well for large separations between fragments (large here means that all fragments are so far that they are almost isolated, $R_{IJ} > 6$ Å). However, for shorter contacts, this model predicts an underestimated screening due to the charge quenching effect (that is, nearby fragments exert a combined potential on a given local area, quenching the charges when the potentials are of the opposite sign).

To improve this local model, a partial model was developed. The main idea of the partial model is to determine partial solvent charges, exerted by each fragment separately, eliminating the charge quenching problem of underestimated screening. To accomplish this, PCM equations have to be solved for each fragment separately. The partial model works quite well for medium and far separations between fragments (for $R_{IJ} > 2$ Å). For far separations, the partial and local models are essentially the same.

As of now, there is no special model for close contacts ($R_{IJ} \leq 2$ Å), and the partial model is the best choice (applied to all pairs). If there are two adjacent charged fragments (R_{IJ} is small), the screening predicted by the partial model does not quite cancel a very strong solute-solute electrostatic interaction.

Now, the important point about screening models is that their usage leaves the total energy invariant. Individual monomer and dimer solvent-related energies are defined differently in each of the two models, which is like shifting furniture in a room, without changing its total weight (the total energy). No matter if partial or local screening is used, the total energy remains the same. Therefore, when doing geometry optimizations or MD, the local screening is better, because it is faster (the local screening is the default). However, to average interactions between fragments directly in MD, the more expensive partial model can be used.

The partial screening model is selected by adding 8 to MODPAR in $PCM, otherwise the local model is used. To accelerate one-body PCM/SMD, 64 can be added to MODPAR in $PCM, and to accelerate both partial and local screening models, 8192 can be added to MODPAR in $FMOPRP. Thus, the recommended options are $PCM MODPAR = 73 (partial) or 65 (local), and MODPAR = 8205 in $FMOPRP for both (here, 8205 = 8192 + 1 + 4 + 8, a combination of 4 suboptions, see Table 20), whereas 73 = 1 + 8 + 64, a combination of 3 suboptions.

In PA (Section 3.5.4), the partial screening model should always be employed, because it is the only model that can be meaningfully used in PA on the grounds of its methodology.

3.4 Polarization, Interaction, and Binding Energies

Absolute QM energies are huge (about 50,000 kcal/mol just for 3 atoms in H_2O), and physically an energy difference is more relevant than absolute values. A difference is computed with respect to some reference state. Understanding what the reference state is and choosing the right reference are of paramount importance, whereas misunderstanding the reference leads to confusion and misconceptions in comparing computed values to other studies.

To give a chemical example, a reaction barrier in a chemical reaction can be computed with respect to the reference state of the reactants. However, reactants often form a complex with some binding energy, so there are two reference states: isolated reactants or bound reactants. The barriers computed for these two reference states differ in the complex binding energy.

In unfragmented calculations, a reference state is defined in terms of the whole isolated molecules. If a protein and a ligand form a complex, the reference state can be defined as an isolated noninteracting protein and ligand (in FMO the protein is divided into fragments, so the reference of each fragment is computed in the protein embedding but without the ligand). FMO also offers a different reference, that of separate isolated fragments (Section 5.1.1). It is important to define the reference clearly in any analysis.

Polarization can be defined as a charge redistribution caused by an external factor, typically, by an embedding ESP. Here, redistribution means that polarization is a change from one distribution to another, that is, between two electronic states. Polarization can be viewed as intrafragment charge transfer.

3.4.1 *Reference States (0, PL0, PL, and CT)*

The physical ideas behind 0, PL0, and PL states (Table 31) come from the energy decomposition analysis,[20] developed about 20 years before FMO, whereas the concept of the CT state is shared in various two-step[3] fragment-based methods.

The non-interacting isolated state, usually denoted by 0, is defined as the name implies. For a water cluster, each water molecule is computed separately. The sum of the energies of isolated fragments is the energy of the isolated state.

The geometry in each state (0, PL0, and PL) may be different. It is possible to define the deformation energy, describing the geometrical

Table 31.   Reference states.

State	Description
0	An isolated, non-interacting state.
PL0	Polarization by the fixed embedding from the 0 state.
PL[a]	Self-consistent polarization (integer charges).
CT	Self-consistent polarization with charge transfer (fractional charges).

[a]PL is the most common state in FMO.

change from one state to another. Here, let us focus on the electronic changes (using the same geometry for different states).

Turning on the embedding in isolated fragments, one obtains PL0 and PL states (PL stands for polarization, induced by the embedding). In the PL0 state, the embedding is frozen (non-polarizable), computed from the electronic states of isolated fragments. The difference between the PL0 and 0 states describes the one-body polarization, that is, the polarization of each fragment is computed independently.

Next, the embedding is relaxed to be polarizable in the PL state, where by repeating fragment calculations self-consistently, the embedding is updated based on the current electronic state of fragments. The difference between PL and PL0 states is the polarization coupling, arising due to the many-body polarization.

The 0, PL0, and PL states deal with fragments assumed to have a fixed integer number of electrons (for example, 10 for an H_2O fragment). Polarization is a redistribution of the electron density in a fragment under the constraint that the total electron count is fixed. This restriction is lifted in the CT state, and electrons are allowed to flow between fragments, that is, the electronic state reflects CT.

There is, however, an obstacle to the use of the CT state in most QM methods, that a wave function can only be constructed for states with a fixed integer number of electrons. So far, the CT state has not been designed for fragments. Rather, it has been used for units called segments (Section 3.5.1), only definable in DFTB, which is based on the physical model conducive to such a definition.

The energy difference for a complex with respect to the non-interacting (0) state is called the binding energy. The energy difference with respect to an embedded (polarized) state is called the interaction energy. In some experiments, a binding energy can be measured. The interaction energy may be a theoretical abstraction, which is directly related to the stability of a complex.

For computational methods neglecting polarization, in vacuum there is no difference between the binding and interaction energies, which may be a reason for some confusion of the two concepts. The relation of the interaction and binding energies is clearly defined in Section 3.5.8.

In solution, the formation of a complex AB is accompanied by a change in the solvation of A and B. That is, some solvent molecules have to go away from the surface of A and B, leaving space for a close contact of A and B, known as a partial desolvation. The binding energy in solution is a combined effect of the polarization, interaction, and desolvation (Section 3.5.8).

The difference between interaction and binding energies can be huge. An interaction energy in a protein-ligand complex can be as large as 100 kcal/mol or more (for charged species), whereas binding energies can be ~10 times smaller, due to a cancellation of the interaction and desolvation energies.

3.4.2 *Polarization of Fragments and Subsystems*

The monomer energies E'_I in FMO are absolute energies, reflecting the destabilization of monomers due to an embedding. Monomer energies can be used to define the destabilization component of the polarization, for which it is necessary to do separate calculations of the 0 and PL0 states. The polarization energy is a sum of the mutual destabilization of monomers (kicked out of their isolated comfort zones) and their stabilization (a gain in the electrostatic energy, which in FMO is included in pair interaction energies).

For a subsystem in a protein-ligand complex (e.g., a protein), one can define the polarization of each residue as the difference in the energies of two PL states: without ligand and with ligand. This gives the polarization energy due to the ligand binding, as opposed to the polarization energy of one residue by all others. This macromolecular view is conveniently handled with the analysis developed for it (Section 3.5.8).

3.4.3 *Pair Interactions (PIE)*

An interaction energy ΔE_{IJ} in the FMO2 expansion in Eq. (2) is a gain or loss in the energy of dimer IJ with respect to the sum of the energies of polarized fragments I and J. In FMO3, there are additional interactions appearing in the triples of fragments (trimers). A three-body term ΔE_{IJK} can be used as a correction to pair interactions, by adding 1/3 of it to each of the three pair interactions (Eq. (18)). These three-body corrected pair interactions $\Delta \tilde{E}_{IJ}$ are a more accurate representation of interactions.

A PIE is not a binding energy. For example, a PIE representing a hydrogen bond is measured in FMO with respect to two polarized fragments, whose energies include the destabilization part of the polarization. The interaction energy includes the stabilization part of the polarization, so that interaction energies are typically more attractive compared to binding energies (another source of a difference in them is desolvation). In describing binding it is necessary to consider the contribution of monomers (the destabilization polarization) to avoid overestimation of energies.

When two fragments are connected by a covalent bond, the interaction energy for such pair is large, on the order of 15 hartree for a C–C bond. This is because of the separation of electrons in the BDA, with some interactions not accounted for in separate monomer calculations, but present in the dimer, so that the difference in the dimer and monomer energies is that large. There is a relatively simple way of separating the artificial BDA interaction from the normal interaction between other fragments, which can be done in PIEDA (Section 3.5.5).

Pair interactions with PBC can be computed in DFTB for fragments or segments (see Section 3.5.1). A PIE/PBC, and its electrostatic and dispersion components, contain an implicit sum over all cells $\mathbf{R}$ (via the Ewald summation).

$$\Delta E_{IJ}\Big|_{I\in 0, J\in 0} = \sum_{R=0}^{\infty} \Delta E_{IJ}\Big|_{I\in 0, J\in R} \tag{25}$$

Pair interactions for two fragments in the original cell **0** are obtained as a sum over cells **R**, including **0** and all replicated cells.

Consider a protein crystal. A residue fragment *I* interacts with all other fragments *J* in the protein in the primary cell **0**; however, they are also replicated cells in the crystal, all of which contain images of *J*, and these fragments *J* ∈ **R** interact with *I*. The sum over cells **R** is ΔE_{IJ}. The interaction of *I* with its own images is included in the monomer energy E_I'.

There is a complication. For protein crystals there may be several copies of the protein per cell, because a space symmetry is destroyed by solvent embedded in the crystal, and in PBC calculations an enlarged cell is created in P1 space group. Interpretation of PIE/PBC in this case is complicated because there are multiple sets of similar fragment pairs per cell.

3.4.4 *Backbone, Connected and Unconnected Dimers*

Pair interaction energies between covalently connected fragments are very large because they include artefacts of fragmentation. Sometimes it is convenient to divide all pair interactions into two classes: those for connected (C) and unconnected (U) dimers. For a connected dimer *IJ*, the separation R_{IJ} between *I* and *J* is 0, because one atom (BDA) is present in both fragments.

The total FMO2 energy can be split into the energy of the backbone (BB) and the energy of unconnected dimers as

$$E^{FMO2} = \sum_{I=1}^{N} E_I^{BB} + \sum_{\substack{I>J: \\ R_{IJ}\neq 0}} \Delta E_{IJ} \tag{26}$$

The backbone energy includes the energy of connected dimers,

$$E_I^{BB} = E_I' + \frac{1}{2}\sum_{\substack{I>J: \\ R_{IJ}=0}} \Delta E_{IJ} \tag{27}$$

where 1/2 is used to avoid double counting, similar to partial energies in Eq. (17).

3.5 Analyses for FMO

In an FMOn calculation ($n > 1$), pair interaction energies between fragments are a by-product of the MBE. Most PIE analyses in FMO are developed for fragments, but some are for segments. It should be explained what the difference is.

3.5.1 *Segments Versus Fragments*

FMO is a method based on QM calculations of fragments. The choice of defining fragments is governed by the consideration of obtaining good accuracy for total properties, such as total energies. The fragments obtained in this way may not be the best units for discussing chemical properties.

It is notoriously difficult to detach peptide bonds, so the fragmentation of proteins in FMO detaches C_α–C bonds and produces fragments that differ from residues by CO groups. It is cumbersome to discuss the contributions of residues and compare to other studies, because fragments in FMO are different from conventional residues.

Another example is a metal ion that forms dative bonds with ligands. To increase the accuracy, the ion and all its chelated ligands are usually put into one fragment (Section 4.11.5). However, it may be desirable to calculate the value of ion-ligand interactions, impossible in FMO because they are all in the same fragment.

In protein-ligand binding, it may be valuable to discuss contributions of functional groups in a ligand, and use that information for improving the efficacy of a drug.

A solution to all of these problems is to perform an FMO calculation with fragments defined in a way to minimize the total error, and then, as a post-processing step, to redefine FMO properties in terms of different units. These units are called segments.

There are essentially no restrictions on defining segments. They can be functional groups, conventional residues, or metal ions.

However, not all FMO calculations can be reprocessed in terms of segments. At present, there are only two tasks where segments may be used. First, the energy in an FMO-DFTB calculation can be divided into segment properties in the partition analysis (PA). Second, vibrational energies in any FMO calculation can be divided between segments in the partition analysis of vibrational energy (PAVE).

3.5.2 *Segments for Electronic Energy Decomposition*

Segments differ from fragments in many important ways, not just in terms of atomic composition. Fragments can have a covalent boundary between them, for example, when a C_α–C bond is detached between residue fragments. The detachment is asymmetric, with a heterolytic division of electrons, 0 $\bar{e}$ to BDA and 2 $\bar{e}$ to BAA (Section 4.11). At a covalent boundary, some special treatment in QM calculations is performed (HOP or AFO). In contrast, there is no treatment of a boundary between segments at all. Each atom in a segment inherits properties from fragment calculations, such as atomic charges, which are used to compute segments. Nothing special is done for segment boundaries; bonds of any order, covalent or dative, may exist between segments.

Such treatment of segments for an energy decomposition is possible owing to the structure of the DFTB energy expression, which permits a natural decomposition into atoms (combined to segments). The electrostatic term in the energy expression is obtained from atomic charges giving rise to a very important conceptual difference between fragments and segments, of a profound nature, both a boon and a curse in some sense. Fragments in FMO have an integer number of electrons assigned to them, and therefore have integer charges.

Segments, on the other hand, use charges of atoms reflecting a CT between them, obtained in FMO2 or FMO3 calculations. Summing atomic charges in a segment gives in general a fractional number. This is a very important conceptual difference to fragments.

Consider a calcium dication. When computed as a fragment in FMO, it has a charge of + 2. The interaction of this + 2 charge with other fragments can be large. Calcium can form dative bonds to some ligands and draw electron density from them, resulting in a fractional charge, perhaps + 0.6. This charge is used to compute properties of Ca as a segment. The formal charge of Ca (+2) used in its fragment is not physical, whereas the fractional charge is more plausible. Interactions of Ca as a segment are usually smaller compared to a Ca fragment, because of a smaller charge. For fragments, the only way to get a better description of such a highly charged fragment prone to a large CT is to use FMO3, with a higher order treatment of CT.

Segments are charged in general. For phenol PhOH, two segments, Ph and OH, can be defined. Each of them has a fractional charge, whereas phenol is neutral as a fragment. If a water molecule interacts with neutral PhOH, the interaction may be small. But for two charged segments, $Ph^{+\delta}$ and $OH^{-\delta}$, there may be two substantial interactions of the opposite sign. In solution, however, solvent screening can decrease the contribution of electrostatics.

Using segments has another advantage over fragments. As discussed above (Section 3.4), the interaction energy between two fragments connected by a covalent bond is very large. In contrast, the interaction energy between two segments is on par with non-covalent interactions, because segments are divided atom-wise compared to the particle-wise division of an atom at a fragment boundary.

For segments, CT is incorporated in the monomer state (the CT state), so that in dimer terms a higher order CT is accounted for, with a more accurate electrostatic interaction than for fragments. On the other hand, the 0-order term in DFTB, that represents exchange and correlation in DFT, is exactly the same for segments and fragments.

Another interesting aspect of using segments is that the MBE of the energy in terms of segments extends to two-body terms only, even if based on the three-body FMO3-DFTB. This is because the energy expression in DFTB has no three-atom potentials; and the total energy for segments has only monomer and dimer terms.

Segment analyses can be applied to a full DFTB calculation without fragmentation. The atomic charges from this calculation can be used to define the energies of segments (technically, such a calculation is accomplished by placing all atoms into 1 fragment).

3.5.3 *Segments for Vibrational Energy Decomposition*

Segments can be used for a vibrational energy decomposition by computing a Hessian, diagonalizing it, and decomposing vibrational energies into segment contributions. Any QM method can be used with PAVE if it has an analytic gradient or Hessian. PAVE can also be used without fragmentation (by defining all atoms to be in 1 fragment).

The technical aspects of this type of calculation are described in Section 3.5.7. PAVE can be used to add vibrational corrections to binding or reaction energies, and pinpoint the quantitative contributions of individual segments. For instance, one can compute how much a hydroxyl group in a ligand contributes to the binding to a protein, and use this information for rational drug design.

3.5.4 *How to Define Segments*

The simplest is to define segments identically to fragments. As discussed above, the energy decomposition for segments differs from the energy for fragments, due to a different treatment of CT.

There are two other ways to define segments. One is to use INDATP in $FMO. The format of INDATP is the same as of INDAT (see Section 2.6.2.1). The other option is to provide PDB data in $PDB. Atomic coordinates in $PDB are always ignored, and $PDB is used only for reading names of atoms, names of segments, and segment IDs (residue sequence numbers). Only ATOM and HETATM records are used, and other records are skipped. The order of atoms in $PDB and $FMOXYZ should be the same. Serial numbers of individual atoms are ignored, so that TER records with an allotted atomic number are not a problem.

The segment ID is read from the column where the residue number is stored. It is allowed to skip some numbers, for example, to start numbering from 127. Internally, segments are always numbered from 1 (the user provided IDs in $PDB are retained in the segment names). All atoms with a given segment ID are placed in the same segment.

Table 32.   Using $PDB to define segments.

Input	Comments
$FMOPRP MONPAN = 1 $END	Perform partition analysis.
$FMO INDATP(1) = –10 $END	A negative number means that $PDB should be read. 10 is the total number of segments (it is permissible to give an overestimate if one is not exactly sure of the number).

Names of atoms in $PDB are only used for printing their labels and also to split residues (optional).

For the PDB way of defining segments, it is necessary to specify their number with a minus as INDATP(1). An example of the input section is shown in Table 32.

It is possible to split side chains from the backbone for some residues. This is done either by adding 32 to MODPAN for splitting all amino acid segments, or by adding 1 to NOPSEG(*i*) for splitting individual segments *i*. Glycines are not affected by the side chain splitting. Termini in a protein may have unexpected leftovers after an automatic splitting of side chains.

If a $PDB is used and the number of segments is specified with a minus in INDATP(1), this number should refer to the total number of segments after the splitting.

The user has full control for defining segments in $PDB by manually changing the segment ID. For example, to split COO⁻ from residue 10, the segment IDs for those 3 atoms can be modified from 10 to 11 (Table 33). 11 is used because the system has 10 residues in total, so the next unused index is 11 (any other unused number can be used such as 12 or 100). As a result, there will be 11 segments for 10 residues. The charge on the COO⁻ segment is automatically calculated, so there is no need to define it (and it is not –1 either; the segment charge is equal to the sum of its atomic charges).

The definition of segments for both electronic and vibrational energy analyses is the same. When vibrational energy is decomposed

Table 33.   Splitting a residue into multiple segments using PDB data in $PDB.

Field	Atom #	Name	res_name	res_ID	x	y	z		
ATOM	131	N	GLY A	10	−1.844	−3.461	3.109	1.00	0.00
ATOM	132	CA	GLY A	10	−1.229	−3.774	4.431	1.00	0.00
ATOM	133	C	GLY A	11ᵃ	−1.877	−5.031	5.014	1.00	0.00
ATOM	134	O	GLY A	11ᵃ	−1.289	−6.092	4.882	1.00	0.00
ATOM	135	OXT	GLY A	11ᵃ	−2.949	−4.912	5.584	1.00	0.00
ATOM	136	H	GLY A	10	−2.159	−2.553	2.918	1.00	0.00
ATOM	137	HA2	GLY A	10	−0.169	−3.939	4.305	1.00	0.00
ATOM	138	HA3	GLY A	10	−1.387	−2.946	5.106	1.00	0.00

[a]Three atoms 133–135 are manually split into a separate segment 11 (COO⁻). If there is a segment 11 elsewhere, then these atoms are merged into it.

in PAVE, the electronic analysis (PA) is not performed (only one analysis can be done at a time).

3.5.5 *Energy Decomposition Analyses for Fragments (PIEDA and EDA)*

Two-body (pair) interaction energies ΔE_{IJ} appear in the MBE of the energy (Section 2.1.1). In FMO3, there are also three-body interactions ΔE_{IJK}, which are three-body interactions in a trimer *IJK* excluding three pairwise *IJ*, *IK*, and *JK* interactions. ΔE_{IJK} is a term representing purely three-body interactions that are usually much smaller than two-body values. Two and three-body interaction energies are examples of many-body interaction energies (MBIE).

To gain a deeper physicochemical insight into interactions, it is useful to decompose MBIEs into components. The natural framework for such a decomposition is the energy decomposition analysis (EDA), originally formulated for full unfragmented calculations.[20]

EDA interfaced with FMO2 is known as the pair interaction energy decomposition analysis (PIEDA), also denoted by FMO2/EDA (for short, EDA2), and EDA combined with FMO3 is denoted as FMO3/EDA (EDA3 in this book for short). One could also refer to the polarization calculations for the PL0 state as FMO1/EDA (EDA1) for completeness (see Section 5.1).

3.5.5.1 *Energy decomposition in PIEDA*

It is natural that an energy decomposition is wave function dependent, because a decomposition is based on terms determined by the underlying physics, and they are tailored to each QM method.

Taking MP2 as an example, a PIE is decomposed as:

$$\Delta E_{IJ} = \Delta E_{IJ}^{ES} + \Delta E_{IJ}^{EX} + \Delta E_{IJ}^{CT+MIX} + \Delta E_{IJ}^{DI+RC} + \Delta E_{IJ}^{BS} + \Delta E_{IJ}^{solv} \quad (28)$$

A summary for different methods is provided in Table 34 (the physical meaning of each term is discussed in greater detail in Ref. 8). The electrostatic (ES) term is the Coulomb interaction. For all methods except DFTB, it is computed between the electron densities and nuclei of two fragments. In DFTB, point charges are used with a damping. The electrostatic interaction is always computed in vacuum (corresponding to the dielectric constant $\varepsilon = 1$). ES can be both attractive and repulsive, and for charged fragments it can be quite large (~100 kcal/mol). ES is a long-range effect.

The charge transfer plus coupling (mix) terms (CT + MIX) describes the effects of CT between two fragments, and some couplings (MIX) of CT to other terms like EX. In DFTB, there is a coupling of charge transfer (CT) to the electrostatic embedding (ES), without any MIX terms (CT·ES). CT + MIX is usually dominated by CT.

For all methods (except DFTB), there is a separate exchange-repulsion (EX) term, which corresponds to the Pauli repulsion between fermions (electrons), related to the antisymmetrizer in the

Table 34.   PIE (ΔE_{IJ}) components in EDA2 (PIEDA) for different QM methods.

Method	Electrostatic	Charge Transfer	Exchange-Repulsion	Remainder Correlation	Dispersion	Basis Set	Solvent
HF	ΔE_{IJ}^{ES}	ΔE_{IJ}^{CT+MIX}	ΔE_{IJ}^{EX}		ΔE_{IJ}^{DI}	ΔE_{IJ}^{BS}	ΔE_{IJ}^{solv}
MP2/CC	ΔE_{IJ}^{ES}	ΔE_{IJ}^{CT+MIX}	ΔE_{IJ}^{EX}	ΔE_{IJ}^{DI+RC}		ΔE_{IJ}^{BS}	ΔE_{IJ}^{solv}
DFT	ΔE_{IJ}^{ES}	ΔE_{IJ}^{CT+MIX}	ΔE_{IJ}^{EX}	ΔE_{IJ}^{RC}	ΔE_{IJ}^{DI}	ΔE_{IJ}^{BS}	ΔE_{IJ}^{solv}
DFTB	ΔE_{IJ}^{ES}	$\Delta E_{IJ}^{CT \cdot ES}$	ΔE_{IJ}^{0+REP}		ΔE_{IJ}^{DI}		ΔE_{IJ}^{solv}

wave function (describing an exchange of two particles). It is a short-ranged QM effect that prevents two opposite charges from collapsing into each other. EX is repulsive and, when this term is not compensated by other attractive terms, it is a probable indication of a steric repulsion in a poorly prepared structure.

For DFTB, ΔE_{IJ}^{0+REP} is the sum of the 0-order (exchange-repulsion and remainder electron correlation), repulsion (REP) energies, and mix terms. The REP energy is extremely short-ranged and has a non-zero value only when two atoms are connected by a covalent bond (in connected dimers). ΔE_{IJ}^{0+REP} can be easily split into ΔE_{IJ}^{0} and ΔE_{IJ}^{REP} (the splitting is usually not done because ΔE_{IJ}^{REP} is essentially zero for all interesting dimers).

The electron correlation, as computed in MP2 or CC, can be classified into two types: dispersion interaction (DI, between instantaneous dipoles of each fragment) and the remainder correlation (RC), which has different contributions including the coupling of an instantaneous dipole in one fragment with CT between fragments.[8] The electron correlation in DFT (excluding double hybrid functionals) is thought not to describe dispersion explicitly (at least, separably), although some dispersion energy may be implicitly taken into account via parametrization. The correlation in DFT is classified as RC.

Dispersion can be added to HF, DFT, and DFTB with parametrized models, the most recent one in GAMESS being D3(BJ). These models are not polarizable, and each chemical element has predefined parameters, fitted for a DFT functional and basis set.

The counterpoise (CP) correction to reduce the basis set superposition error (BSSE) is not used in PIEDA. There are two ways to account for basis set effects. One is via HF-3c (Section 4.1.2), where the BS component is parametrized, like dispersion. The other is to use AP (Section 5.4), where the BS term is computed as the difference in PIEs for two basis sets, a medium and a large one. DFTB has a reduced BSSE because core electrons are not present. There is no way to add an explicit basis set correction in DFTB.

The solvent screening (Section 3.3) is available in PCM or SMD. It is computed from the interaction of solute charge distributions with induced solvent charges.

3.5.5.2 *Energy decomposition in EDA3*

In FMO3-DFT, the three-body interaction energy can be decomposed as

$$\Delta E_{IJK} = \Delta E_{IJK}^{EX} + \Delta E_{IJK}^{CT+MIX} + \Delta E_{IJK}^{RC} + \Delta E_{IJK}^{BS} + \Delta E_{IJK}^{solv} \qquad (29)$$

A summary of components is provided in Table 35. It is a shorter set compared to two-body terms in Table 34. This is because some contributions, namely, ES, empirical DI and BS, and REP are pairwise additive, so that three-body terms are exactly zero. Individual values of MBIEs are usually small, rarely exceeding 1–2 kcal/mol, in which case it is an indication of a loss of FMO2 accuracy.

ΔE_{IJK} can be contracted with ΔE_{IJ}, producing three-body corrected PIEs $\Delta \tilde{E}_{IJ}$ in Eq. (18). They can be decomposed as regular two-body PIEs.

3.5.5.3 *Input files and an application example*

Most commonly, EDA*n* calculations are performed for the PL state (Section 3.4.1) by simply setting IPIEDA = 1 in $FMO. On the other hand, IPIEDA = 2 is used to compute polarization (Section 5.1.1) in a multistep series of calculations.

For FMO/EDA, RUNTYP = ENERGY in $CONTRL and MODORB = 3 in $FMOPRP should be used. Some wave functions, for example

Table 35.   MBIE (ΔE_{IJK}) components in EDA3 for different QM methods.

Method	Charge Transfer	Exchange-Repulsion	Remainder Correlation	Dispersion	Basis Set (AP)	Solvent
HF	ΔE_{IJK}^{CT+MIX}	ΔE_{IJK}^{EX}			ΔE_{IJK}^{BS}	ΔE_{IJK}^{solv}
MP2/CC	ΔE_{IJK}^{CT+MIX}	ΔE_{IJK}^{EX}		ΔE_{IJK}^{DI+RC}	ΔE_{IJK}^{BS}	ΔE_{IJK}^{solv}
DFT	ΔE_{IJK}^{CT+MIX}	ΔE_{IJK}^{EX}	ΔE_{IJK}^{RC}		ΔE_{IJK}^{BS}	ΔE_{IJK}^{solv}
DFTB	ΔE_{IJK}^{CT-ES}		ΔE_{IJK}^{0}			ΔE_{IJK}^{solv}

MCSCF, may not be used in PIEDA. PIEDA may be combined with SA (Section 3.5.8) and PA (Section 3.5.6), but not with FA.

In a demonstrative application, PIEDA is applied to phenol in explicit solvent (8 water molecules) shown in Figure 14a. The system is surrounded by implicit solvent (PCM). Water molecules solvate the water-loving (hydrophilic) hydroxyl group and befriend other water, eschewing the water-fearing (hydrophobic) phenyl ring. Phenol is fragment $I = 1$, and water molecules are numbered from 2 to 9.

The matrix of ΔE_{IJ} is visualized in Figure 14b as a 2D heat map. Because there are no self-interactions, the diagonal $I = J$ is zero. There are very few slightly repulsive (positive) water-water interactions, for example, between fragments 8 and 4, and the rest are attractive (negative). Water-water hydrogen bonds in water are about −5 kcal/mol.

It is easy to see which water molecules interact strongly with phenol by looking at the 1D plot of total PIEs in Figure 14c. Three hydrogen bonds are formed between phenol and water

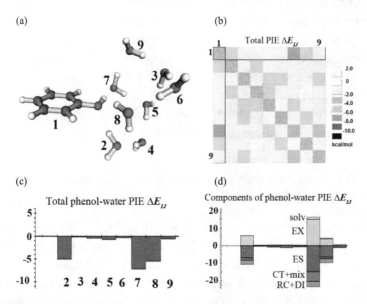

Figure 14. PIEDA/PCM at the level CCSD(T)/aug-cc-pVDZ (RESPPC = −1 is used to accommodate for the diffuse basis, see Section 5.3). a) Structure, b) heat map of all PIEs ΔE_{IJ}, c) solute-solvent PIEs ΔE_{IJ} ($J = 1$), and d) a decomposition of ΔE_{IJ} ($J = 1$) in PIEDA.

(fragments 2, 7, and 8), with the interaction energy from −5.1 to −7.4 kcal/mol. The strongest is with fragment 7, because the 3 atoms in OH...H are able to adopt a favorable linear conformation.

The PIEDA components are shown in Figure 14d. The ES term is the main contribution to hydrogen bonding (as much as −14.8 kcal/mol for the 7,1 pair). EX is usually repulsive, at most 15.0 kcal/mol in this system; it counterbalances the ES term. The stronger the binding, the more repulsive the EX term is, because an attractive interaction decreases the separation increasing a short-range repulsion. The other repulsive term is the solvent screening (solv), which, however, is relatively small (up to 1.4 kcal/mol) in this system. Solvent screening usually reduces ES interaction in the solute.

The charge transfer term (CT + MIX) is attractive, up to −5.7 kcal/mol. The electron correlation (RC + DI) contributes up to −3.2 kcal/mol. In the 7,1 pair, this term is larger than in the other two phenol-water pairs by about 1 kcal/mol. This extra gain in dispersion may come from the interaction of the delocalized π electrons in the benzene ring with the lone pairs of water, which lean toward each other.

In general, EX and CT + MIX components are substantial only for close contacts. When a structure is well optimized, the repulsion (EX) is less than the attraction (ES + solv, CT + MIX, and DI). However, when several fragment pairs interact, it is often inevitable that in order to have a strong attraction for one pair, another pair may be repulsive (sacrifice a little to gain more).

3.5.6 *Partition Analysis for Interactions Between Segments (PA)*

Partition analysis (PA) is a segment analogue of PIEDA. PA is based on segments (Section 3.5.1) and CT state (Section 3.4.1). PA can be conducted for full unfragmented calculations, technically accomplished as an FMO calculation with 1 fragment. PA is a post-processing scheme, which uses QM calculation results (mainly atomic charges), without performing new segment-specific QM calculations. At present, PA can be done for DFTB2 or DFTB3, with a solvent model (PCM or SMD), dispersion, and PBC. PA can be used with the partial screening model (Section 3.3).

PA begins after an FMO-DFTB calculation finishes, taking atomic charges from FMO, and summing them to get segment charges Q_i according to the definition of segments. Then, individual segment properties are computed: internal energies E'_i, solvent-related energies ΔE_i^{solv}, and induced solvent charges q_i.

The total energy for M segments looks like an FMO2 total energy,

$$ E = \sum_{i=1}^{M} \left(E'_i + \Delta E_i^{solv} \right) + \sum_{i>j} \Delta E_{ij} \qquad (30) $$

An important difference to fragments is that the energy expression for segments stops at the two-body level, because all terms in DFTB are of at most two-particle nature. Thus, no matter if FMO2, FMO3, or full DFTB charges are used to define segments, the expression in Eq. (30) stays the same.

An important feature of PA is that the total energy E in Eq. (30) is exact, when exact charges are used for segments. When FMO charges are used, the total energy for segments is different from the total energy for fragments, because the CT treatment is different: PA includes higher-order coupling terms (four-body for FMO2 and six-body for FMO3) than FMO. The treatment of high-order terms, however, is partial, and the energy in PA is not always more accurate than the energy in FMO, although it often is.

The celebrated property of PA that the energy E in Eq. (30) is invariant to a segmentation. Provided that the same "input" atomic charges Q_A are used, the energy E stays the same no matter how many segments there are and how they are defined. There is only one known condition that slightly violates the invariance — the escaped charge compensation ICOMP = 2 in PCM, because with this feature the induced solvent charges are renormalized for each segment, and thus the charge renormalization depends on the segmentation, albeit the energy is weakly affected by that.

Zero-order interactions (corresponding to exchange-correlation in DFT) are included only in monomer terms without any contributions in PIEs, whereas CT is integrated into ES. Segment energies are decomposed as

$$E_i' = E_i^0 + E_i^{ES} + E_i^{REP} + E_i^{DI} \tag{31}$$

Pair interaction energies in PA can be decomposed as

$$\Delta E_{ij} = \Delta E_{ij}^{ES} + \Delta E_{ij}^{REP} + \Delta E_{ij}^{DI} + \Delta E_{ij}^{solv} \tag{32}$$

where the electrostatic (ES) term has a similar meaning as in PIEDA, but it is computed for segments with fractional charges. The repulsion (REP) energy is a very short-ranged term that is essentially zero except for covalently connected segments, in which case it is several kcal/mol per bond. The repulsion energy is an extra fitting term in DFTB, evasive in its physical meaning. The dispersion (DI) term is rather like in PIEDA for DFTB.

The solvent term is a sum of electrostatic (es) and non-electrostatic (non-es) screenings. The latter is found only in PCM, the sum of disp and rep terms.

$$\Delta E_{ij}^{solv} = \Delta E_{ij}^{es} + \Delta E_{ij}^{non-es} \tag{33}$$
$$\Delta E_{ij}^{non-es} = \Delta E_{ij}^{disp} + \Delta E_{ij}^{rep} \tag{34}$$

An interesting aspect of using segments is that there is no redundancy of atoms at a segment boundary, and PIEs for segments connected by a covalent bond are on the same order of magnitude as non-covalent interactions. This is in part a result of the distribution of the zero-order energy E_i^0 in DFTB among segments in such a way that only monomers get a contribution, but not dimers.

There is no limit other than your imagination in defining segments. A segment may be an atom (for example, a metal cation), a functional group (such as –OH, –COO⁻, etc.) or a real residue in a protein, all of which are rather impossible for fragments.

To do a PA calculation, one has to define segments (Section 3.5.4). There is one option MODPAN (Table 36) for defining global details of PA, and an array to define options for each segment

Table 36. Packed MODPAN option in $FMOPRP for defining details of PA.

Value	Meaning
1	Turn PA on.
2	Split the non-es term into the rep and disp terms in PCM.
8	Split all residues into backbones and side chains.
16	Split REP into monomer and dimer values.
32	Set segments to be equal to fragments (no need to define INDATP).

NOPSEG (at present, the only choice is to set 1 for residues in which one wishes to split the side chain from the backbone). For example, NOPSEG(5) = 1,0,1 will result in residues 5 and 7 being split into two segments each, which requires that $PDB should be provided in the input.

When PCM is used, MODPAN = 2 can be set to split the non-electrostatic screening non-es into disp and rep terms (because both are usually small, they are not split by default).

The full decomposition in Eqs. (31) and (32) is only used when MODPAN = 16 is set. By default, however, the MODPAN = 16 feature is not used, and REP terms are compressed into partial energies $\tilde{E}_i^{REP}$, using Eq. (17), applied to REP energies. Thus, the most commonly used PA equations are

$$E_i' = E_i^0 + E_i^{ES} + \tilde{E}_i^{REP} + E_i^{DI} \tag{35}$$

$$\Delta E_{ij} = \Delta E_{ij}^{ES} + \Delta E_{ij}^{DI} + \Delta E_{ij}^{solv} \tag{36}$$

The main cost of a PA calculation is to solve for PCM charges in Eq. (22) for each segment. This step is parallelized similar to monomers in FMO, using NGRFMO(1) groups. For PA executed with GDDI, it is required that a group should have 1 core, so that the number of groups should be equal to the number of cores.

3.5.7 *Partition Analysis for Vibrations (PAVE)*

Using harmonic frequencies in thermodynamical partition functions, it is easy to calculate vibrational complements to electronic energies.

PAVE can use both analytic and semi-analytic Hessians. A PAVE calculation can only be done with RUNTYP = FMOHESS.

To use PAVE, one should define a set of active atoms. A semi-numerical Hessian is built by doing numerical shifts of these active atoms. In the case of analytic Hessians, first a full Hessian is computed and then the block for the active atoms is extracted. In either case, the Hessian for active atoms is diagonalized, and its eigenvalues and eigenvectors are obtained. It is permissible to use all atoms as active.

The following vibrational quantities are decomposed into segment contributions: zero point energy (ZPE) E^{ZPE}, temperature-dependent contribution H^T to the enthalpy H, entropy S and free energy G. Enthalpy and internal energy differ by the pressure-volume PV term, which is considered to have no vibrational contribution, thus vibrational enthalpy and internal energy may be considered to be the same. Likewise, Helmholtz and Gibbs free vibrational energies coincide.

The total free energy is decomposed into M segment contributions as

$$G = \sum_{i=1}^{M} G_i \tag{37}$$

Likewise, all other contributions like H and S are decomposed. The relationships between thermodynamical quantities are:

$$H = E^{ZPE} + H^T \tag{38}$$

$$G = H - TS \tag{39}$$

Segments in PAVE may be used for any QM method, not just DFTB. The definition of segments is the same as in PA (Section 3.5.2). Thus, it is possible to compute free energy and entropy for any segment, such as amino acid residues or functional groups.

ZPE is independent of temperature, being mainly determined by high frequency vibrations. Enthalpy has a small temperature-dependent term H^T added to ZPE. Entropy is temperature-dependent, and it is mainly determined by low frequency vibrations.

Low frequency vibrations are problematic due to (a) approxima-tions in the Hessians, (b) an imperfectly found stationary point (the gradient is not exactly 0), (c) violations of rotational invariance (in DFT and PCM), and (d) numerical issues. In general, the more active atoms there are, the smaller low frequency vibrations become. For many active atoms, differences in entropy and free energy may include a large numerical noise.

When using PAVE, one should decide how many low frequencies to exclude from the analysis. When a partial set of atoms is used in PAVE, there are no zero frequencies corresponding to rotational and vibrational degrees of freedom, because if a part of the system is translated or rotated, the total energy can change. Therefore, in prin-ciple all vibrations can be included. Due to a non-zero gradient, inaccuracies in the Hessian, or for a transition state (TS), some vibra-tions may be imaginary, and they are always excluded in PAVE. To reduce numerical issues, it is recommended to optimize to at least OPTTOL = 1e-5.

Consider a complex of a protein and a ligand. The vibrational part of the binding free energy is

$$\Delta G = G_{complex} - G_{protein} - G_{ligand} \qquad (40)$$

When a non-linear ligand is computed as a standalone molecule, 6 frequencies are zero, and to match that, one may like to also exclude 6 low frequencies in the complex, although any number of vibrations may be excluded (NZPHA in $FORCE). Combining Eqs. (37) and (40), vibrational contributions of segments to the total bind-ing can be obtained.

For PAVE, an input file for a Hessian has to be created. Active atoms are specified with JACTAT, the format of which is similar to the list version of INDAT: atoms and ranges may be given, for example, JACTAT(1) = 6 11 −13 defines atoms 6, 11, 12, and 13 as active.

For DFTB in vacuum, one can consider using an analytic Hessian; in many other cases it may be better to use a semi-numerical Hessian. Up to 10 temperatures can be used in one run by specifying TEMP in $FORCE. A restart of a Hessian is not possible. IR intensities and

Table 37.   Options for a PAVE calculation.

Input file	Comments
⌐$CONTRL RUNTYP = FMOHESS $END	Semi-numerical Hessian.
⌐$FORCE METHOD = SEMINUM	Shift coordinates by 10^{-3} bohr.
VIBSIZ = 1E-3	Exclude 6 smallest (or imaginary)
NZPHA = 6	frequencies from thermochemistry.
	Atoms 11–13 are active.
JACTAT(1) = 11 –13 $END	Segments = fragments.[a]
⌐$FMOPRP MODPAN = 33 $END	

[a]Alternatively, MODPAN = 1 and INDATP should be given to define segments.

Raman activities may be calculated (Sections 3.9.2 and 3.9.3). An example of a header for a PAVE calculation is given in Table 37.

3.5.8 *Subsystem Analysis for Analyzing Binding (SA)*

Computing an energy change in some process is a very common physicochemical task. Without a loss of generality, such a change can be written in the form of a chemical reaction, which can also describe the binding energy in a complex formation. An example of such a reaction adopted in this Section for the sake of a concrete explanation is

$$A + B \rightarrow AB \tag{41}$$

The binding energy of subsystem A (e.g., a protein) to subsystem B (e.g., a ligand) is computed as the energy difference,

$$\Delta E^{\text{bind}} = E^{AB} - E^A - E^B \tag{42}$$

The geometries of A and B in the complex are in general different from their isolated state. The following expression constitutes the SA,

$$\Delta E^{\text{bind}} = \Delta E'_A + \Delta E'_B + \Delta E^{\text{def},A} + \Delta E^{\text{def},B} + \Delta E^{\text{int}} \tag{43}$$

Here, the deformation (def) energy of A (likewise, B) is defined as the energy difference of A at the geometry of the complex $A|AB$ and in the isolated state.

$$\Delta E^{\text{def},A} = E^{A|AB} - E^A \tag{44}$$

$\Delta E^{\text{def},A}$ shows how much energy has to be spent to deform A so that it can bind to B. For example, a binding pocket may have to open up so that B can enter, which costs energy. B may have to contract its protruding groups for the same purpose ($\Delta E^{\text{def},B}$).

All terms in Eq. (43) except for deformation energies are calculated for the same molecular geometry of AB in the complex.

In solution, $\Delta E'_A$ describes two effects: (1) the destabilization polarization of A, which is the change in the electronic state of A, caused by the presence of B in the complex and (2) the desolvation of A (some solvent has to move away from the surface of A to give space for B to bind). It is possible to separate these two effects.

ΔE^{int} is the interaction energy between A and B, defined by summing PIEs,

$$\Delta E^{\text{int}} = \sum_{I \in A} \sum_{J \in B} \Delta E_{IJ} \tag{45}$$

Summarizing, A and B have to deform to bind each other (deformation energy), divest some solvent (desolvation penalty), undergo an electronic change (polarization), and, finally, interact (interaction energy). The polarization here is taken in the molecular sense, as discussed more in Section 3.4.2. It is clear that the TIE, ΔE^{int}, is just one component of the binding energy, and numerically, ΔE^{int} can be 1–2 orders of magnitude larger than ΔE^{bind} (see also Section 3.4).

The analysis in Eq. (43) is a subsystem-based decomposition. One can stop there, or dive one level deeper by defining fragment contributions using partial energies in Eq. (17). For A (likewise, for B),

$$\Delta E'_A = \sum_{I \in A} \Delta E_I^{\text{part}} \tag{46}$$

using the change in the partial energies of fragments in A due to complexation. ΔE_I^{part} is obtained by subtracting E_I^{part} in the isolated state from the value in the complex, for $I \in A$.

$$\Delta E_I^{part} = E_I^{part,AB} - E_I^{part,A} \tag{47}$$

It is possible to go still deeper, for example, if one is interested in things such as: how much does the interaction energy of Ala-10 with Pro-20 in the protein change because of the ligand polarization? In this case, the FMO2 expansion can be used to obtain the change in ΔE_{IJ} due to complexation, answering the posed question. Such three-body results may be useful to explain the change in the protein function upon activation and allosteric regulation. FMO3 can be used for a further plunge in understanding complex many-body effects.

Finally, it remains to be shown how to split polarization and desolvation. There is some methodological freedom in doing it. The path chosen here is to consider the change in the solvation energies of fragments as the desolvation, whereas the solvent contribution for dimers (screening) is considered together with the solute-solute interactions, that is, as a polarization.

The differential partial energies of fragments ΔE_I^{part} can be easily split into the solute energies ΔE_I^{pPLd}, describing the partial destabilization polarization (pPLd), and differential solvent-related energies $\Delta\Delta E_I^{solv}$, describing the desolvation (the double Δ is used because it is a difference of two solvent-related energies of fragment I, in the bound and isolated states, and each of these two energies has a Δ).

$$\Delta E_I^{part} = \Delta E_I^{pPLd} + \Delta\Delta E_I^{solv} \tag{48}$$

In practice, ΔE_I^{part} is calculated according to Eq. (47) for $I \in A$ (likewise for B). Then, the desolvation penalty for A (likewise for B) is computed as

$$\Delta\Delta E_I^{solv} = \Delta E_I^{solv,AB} - \Delta E_I^{solv,A} \tag{49}$$

and, finally, the polarization contribution is obtained as

$$\Delta E_I^{pPLd} = \Delta E_I^{part} - \Delta\Delta E_I^{solv} \tag{50}$$

Thus, the polarization (ΔE_I^{pPLd}) and desolvation ($\Delta\Delta E_I^{solv}$) energies can be obtained. The stabilization part of the polarization is included in the interaction energy, not easily separable (see Section 5.1 for a discussion of the stabilization and destabilization parts of polarization and note that the polarization here is of one subsystem by another as a whole, whereas Section 5.1 deals with the polarization of one fragment by another in the same system).

Plugging Eq. (46) in Eq. (43), a detailed form of SA is obtained,

$$
\begin{aligned}
\Delta E^{bind} &= \sum_{I\in A}\Delta E_I^{pPLd} + \sum_{I\in B}\Delta E_I^{pPLd} + \sum_{I\in A}\Delta\Delta E_I^{solv} + \sum_{I\in B}\Delta\Delta E_I^{solv} + \Delta E^{def,A}\\
&\quad + \Delta E^{def,B} + \sum_{I\in A}\sum_{J\in B}\Delta E_{IJ}\\
&= \Delta E^{PLd,A} + \Delta E^{PLd,B} + \Delta E^{desolv,A} + \Delta E^{desolv,B} + \Delta E^{def,A} + \Delta E^{def,B}\\
&\quad + \Delta E^{int}
\end{aligned}
\tag{51}
$$

where the binding energy is clearly decomposed into the destabilization polarizations of A and B, desolvation energies of A and B, deformation energies of A and B, and the interaction energy.

When B is not divided into fragments, for example, when B is a ligand, a further useful simplification is possible, obtained from Eqs. (43) and (46), without splitting the partial energies,

$$\Delta E^{bind} = \sum_{I\in A,B}\Delta E_I^{bind} + \Delta E^{def,A} + \Delta E^{def,B} \tag{52}$$

The total binding energy is here decomposed into fragment binding energies. The deformation energies are not split into fragment contributions (because they are usually small), but they can be. Ligand B contributes to binding due to its desolvation and polarization (because B has just 1 fragment $I \in B$) and the binding energy contribution is defined as

$$\Delta E_I^{bind} = \Delta E_I^{part} = \Delta E_I' + \Delta E_I^{solv} \tag{53}$$

The binding energy of residue fragment $J \in A$ is defined to include protein-ligand interaction ΔE_{IJ} of residue J with ligand $I \in B$,

$$\Delta E_J^{bind} = \Delta E_J^{part} + \Delta E_{IJ} \tag{54}$$

A useful way to understand a complex formation in solution is to rewrite[21] Eq. (41) as

$$A \cdot s_A + B \cdot s_B \rightarrow AB + s_A \cdot s_B \tag{55}$$

where it is explicitly shown that prior to binding B, A interacts with solvent s_A (likewise for B). In other words, two complexes are broken and two are created. The view that one complex is formed in Eq. (41) is misleading (in solution). The desolvation penalty is the energy to break $A \cdot s_A$ and $B \cdot s_B$ complexes, and create $s_A \cdot s_B$, computed in PCM for implicit continuum solvent. On the other hand, with explicit solvent, it can be done at the molecular level.[21]

SA has other important usages, which can be described in a manner of a chemical reaction. For example, isomer stability can be analyzed, as in α-helix, β-strand, and other motifs in proteins. In this case, the isomerization reaction is simply $A \rightarrow A'$, so that the energies of two isomers are subtracted and decomposed.

Creating an input for SA is very easy: add MODMOL = 8 to $FMO and define MOLFRG(1) as a subsystem index for each fragment. For example, MOLFRG(1) = 1,1,2,2,1,3,3 defines 3 subsystems, the first containing fragments 1, 2, and 5, the second with fragments 3 and 4, and the third with fragments 6 and 7. In order to define partial energies of fragments, SA calculations have to be conducted not only for the complex AB, but also for isolated subsystems A and B. The latter is done by setting all MOLFRG values to 1. If a subsystem has 1 fragment, then it can be calculated either using FMO1 or as an unfragmented calculation (the total QM energy is equal to the partial energy in this case), with the same result.

The number of subsystems cannot exceed N (the number of fragments). The above analysis is derived for the complex formation in Eq. (41), and it can be easily modified to accommodate other processes.

There is also SA for segments in PA. To use it, the second vector in NOPSEG is set like MOLFRG. Namely, if there are 10 segments, then NOPSEG(1:10) defines segment options, whereas NOPSEG (11:20) defines segment subsystems, so setting NOPSEG(11) = 1,2,2,1,1, 3,3,1,2,1 would define 3 subsystems. MODMOL has no effect on segments. It is possible to do two subsystem analyses in the same PA run, one for fragments and another for segments, by setting both MOLFRG and NOPSEG.

As a word of caution, the choice of computing or ignoring the deformation energy is not an easy decision. On the one hand, it is definitely the right thing to include it; on the other, one can fail to locate global minima in separate optimizations of *A*, *B*, and *AB*. Ignoring the deformation might turn the error cancelation to an advantage.

Performing SA involves several steps, listed in Table 38. When one of the subsystems consists of a single fragment, it can be computed without FMO or as a 1-fragment FMO. Although it is possible to use FMO to optimize a system with 1 fragment (for a ligand), it is better to do a non-FMO calculation for that purpose. At the end of the day, the sum of all terms in a decomposed binding energy should match the value of the total binding energy in Eq. (42), the check sum.

Although SA is general and can be applied as presented for the case when there are detached covalent bonds between subsystems, the interpretation of the interaction energy is hampered by the artefacts of fragment boundaries. Whereas it is possible to use BDA corrections (Section 4.11.3), or the concept of the backbone (Section 3.4.4), it is most straightforward to apply SA to subsystems that are standalone, without covalent bonds between them, for example, to a protein and a ligand, or two proteins. On the other hand, segments have no artefacts of bond detachment, so SA in terms of segments may be applicable to covalently bound subsystems without further complications (for example, to ligands that bind covalently to proteins). SA for segments is formulated in essentially the same way as for fragments.

Table 38. Protocol for doing SA calculations for $A + B \rightarrow AB$.[a]

Step	Results	Description
1. Optimize geometry of A.	E^A	
2. Optimize geometry of B.	E^A	
3. Optimize geometry of AB.	E^{AB}	
4. Extract geometry of A from the optimized complex AB and do an SA calculation with 1 subsystem.	$E_I^{\text{part},A}$ $\Delta E_I^{\text{solv},A}$ $E^{A|AB}$	Set MOLDMOL = 8 and MOLFRG(1) = 1,1,...,1
5. Extract geometry of B from the optimized complex AB and do an SA calculation with 1 subsystem.[b]	$E_I^{\text{part},B}$ $\Delta E_I^{\text{solv},B}$ $E^{B|AB}$	Set MOLDMOL = 8 and MOLFRG(1) = 1,1,...,1 (not 2, as it is the only subsystem).
6. Compute deformation energies of A and B.	$\Delta E^{\text{def},A}$ $\Delta E^{\text{def},B}$	Use Eq. (44).
7. Perform subsystem calculation for complex AB.	$E_I^{\text{part},AB}$ $\Delta E_I^{\text{solv},AB}$	Set MOLDMOL = 8 and MOLFRG with an appropriate combination of 1 and 2.
8. Compute desolvations and polarizations of fragments in A and B.	$\Delta\Delta E_I^{\text{solv}}$ ΔE_I^{pPLd}	Use Eqs. (49) and (50).
9. Calculate the interaction energy.	ΔE^{int}	Use Eq. (45).
10. Obtain the total decomposition.	ΔE^{bind}	Use Eq. (51).

[a]It is possible to ignore the deformation energy and skip steps 1, 2, and 6.

[b]For a 1-fragment calculation of ligand B, there is no embedding and no dimers, so the total energy $E^{B|AB}$ is the same as $E_I^{\text{part},B}$.

In the general case of multiple subsystems, the partial energy of subsystem i can be defined similarly to Eq. (17) as

$$E_i^{\text{part}} = E_i' + \frac{1}{2}\sum_{j \neq i} \Delta E_{ij} \qquad (56)$$

where the internal energy E_i' of subsystem i is defined as the total energy in Eqs. (2) and (3) with the restriction that all fragments appearing in them are in *i*. The interaction energy ΔE_{ij} of subsystems *i* and *j* is defined similar to Eq. (45) for *i* = *A* and *j* = *B*.

PIEs for unconnected (U) dimers are summed over fragments in a subsystem (*j*),

$$\Delta E_{Ij}^{U} = \sum_{\substack{J \in j \\ R_{IJ} \neq 0}} \Delta E_{IJ} \tag{57}$$

By restricting dimers in the sum for ΔE_{ij} to be unconnected ($R_{ij} \neq 0$), ΔE_{ij}^{U} can be defined.

3.5.9 *Fluctuation Analysis for MD (FA)*

Fluctuation analysis (FA) is a tool for averaging energies over a trajectory in MD, by taking energies for some snapshots (governed by JEVERY in $MD). The purpose is to include the effect of temperature on the energy (corresponding to the internal energy in the thermodynamical sense).

This is done by averaging energies over time steps (indicated by brackets). It is useful to define a reference energy E_0 as the lowest energy in MD. Fluctuations can be computed as deviations from E_0.

$$\langle E \rangle = E_0 + \langle E - E_0 \rangle = E_0 + \langle \Delta E \rangle \tag{58}$$

For FMO2,

$$\langle E \rangle = \sum_{I=1}^{N} E_I'^{0} + \sum_{I>J} \Delta E_{IJ}^{0} + \sum_{I=1}^{N} \langle \Delta E_I' \rangle + \sum_{I>J} \langle \Delta \Delta E_{IJ} \rangle \tag{59}$$

In other words, the energy $\langle E \rangle$ averaged in MD is the sum of monomer energies $E_I'^{0}$ and PIEs ΔE_{IJ}^{0} for the reference state, plus fluctuations of monomer energies $\langle \Delta E_I' \rangle$ and PIEs $\langle \Delta \Delta E_{IJ} \rangle$. The geometry of the reference (usually the lowest energy in MD) is included in the printed summary of FA results, as well as the ranges in which various values change (min and max). When FA is applied to FMO3, only FMO2 values are automatically averaged at present.

Another quantity that FA provides is the averaged kinetic energy $\langle E_I^{kin} \rangle$ of each fragment, and the temperature T_I it corresponds to.

In NVT and NPT ensembles, the total kinetic energy is constant (normalized to match the global temperature T), but some fragments are "cold" and some are "hot", as determined by their kinetic energy.

$$T_I = \frac{2\langle E_I^{kin}\rangle}{3N_I^{at}k_B} \tag{60}$$

where k_B is the Boltzmann's constant and N_I^{at} is the number of atoms in fragment I.

3.5.10 *Free Energy Decomposition Analysis (FEDA)*

Free energy and entropy are of major interest in soft materials where their calculation may be essential for a reliable theoretical analysis of binding. For this purpose, FEDA is developed.

To decompose free energy, one has to compute it first. This can be done in FMO-MD simulations. At present, the only practically usable way is to do an umbrella sampling MD (Section 3.10.2), which is a relatively straightforward approach for chemical reactions, but for other processes, such as protein-ligand binding, it may not be feasible.

The practical scheme of FEDA is as follows. First, a reaction coordinate ζ should be chosen, for example, a bond length or an angle. Then US MD simulations are performed for a set of values of ζ_0 on a grid by applying the harmonic constraint potential with a force constant k,

$$U(\zeta) = \frac{k}{2}(\zeta - \zeta_0)^2 \tag{61}$$

From the obtained set of trajectories for all ζ_0, the potential of mean force taken as a measure of the free energy $F(\zeta)$ is calculated using weighted histogram analysis.[22] From a plot of $F(\zeta)$, two values of ζ are chosen, representing a desirable transition, for example, reactants (A) and transition state (B), ζ^A and ζ^B, respectively (Figure 15). Then the free energy change is computed as

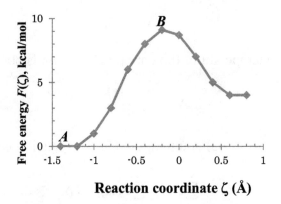

Figure 15. Schematic plot of a reaction profile showing two points of interest, A (reactants) and B (transition state).

$$\Delta F = F(\zeta^B) - F(\zeta^A) \tag{62}$$

Next, the internal energy change is calculated by doing two more MD simulations, constraining the reaction coordinate to the values of $\zeta_0 = \zeta^A$ (reactants) and $\zeta_0 = \zeta^B$ (transition state) with a large force constant k, on the order of 10^5 kcal mol^{-1} U^{-2} (U is Å or °). The total QM energy is averaged and decomposed following FA (Section 3.5.9) but using state A as the reference,

$$\Delta E = \left\langle E\left(\zeta^B\right)\right\rangle - \left\langle E\left(\zeta^A\right)\right\rangle = \sum_{I=1}^{N} \Delta E'_I + \sum_{I>J}^{N} \Delta\Delta E_{IJ} \tag{63}$$

where the change in the energy of fragment I is

$$\Delta E'_I = \langle E'_I(\zeta^B)\rangle - \langle E'_I(\zeta^A)\rangle \tag{64}$$

and the change in the PIE is

$$\Delta\Delta E_{IJ} = \langle \Delta E_{IJ}(\zeta^B)\rangle - \langle \Delta E_{IJ}(\zeta^A)\rangle \tag{65}$$

The Helmholtz free energy ΔF in NVT MD is (to get Gibbs free energy, NPT MD should be done)

$$\Delta F = \Delta E - T\Delta S \tag{66}$$

where T is the temperature. The entropy change ΔS is obtained as

$$\Delta S = \frac{\Delta E - \Delta F}{T} \tag{67}$$

Combining Eqs. (63) and (66), the FEDA expression is obtained

$$\Delta F = \sum_{I=1}^{N} \Delta E_I' + \sum_{I>J}^{N} \Delta \Delta E_{IJ} - T\Delta S \tag{68}$$

The entropy is not decomposed into fragment values, as it is a property of the conformational space of the whole system.

This analysis can be applied to a chemical reaction in explicit solvent (water), where A and B are the reactants and transition state, respectively. One fragment M describes the reactants; $\Delta E_M'$ for this fragment is the direct contribution of the reactants to the reaction barrier, including the effect of the polarization by the solvent. $\Delta E_I'$ ($I \neq M$) for solvent molecules describes the polarization and deformation of the solvent by the reaction fragment. $\Delta \Delta E_{MI}$ is the contribution of the solute-solvent interactions to the reaction barrier. For other pairs (both I and J are solvent), $\Delta \Delta E_{IJ}$ is the change in the solvent-solvent interactions due to the formation of the transition state.

ΔE_I and $\Delta \Delta E_{IJ}$ for water fragments I and J can be combined into a single subsystem energy of water (Section 3.5.8), obtaining the total destabilization polarization of the solvent in the reaction. The barrier $\Delta E^{\ddagger}$ is decomposed into the contributions of the reaction fragment $\Delta E^{\ddagger}_{\text{reactant}}$, solvent polarization + deformation $\Delta E^{\ddagger}_{\text{solvent}}$, and reactant-solvent interactions $\Delta \Delta E^{\ddagger}_{\text{reactant-solvent}}$

$$\Delta E^{\ddagger} = \Delta E_{\text{reactant}} + \Delta E^{\ddagger}_{\text{solvent}} + \Delta \Delta E^{\ddagger}_{\text{reactant-solvent}} \tag{69}$$

which is convenient, because considering contributions of individual solvent molecules in MD may be an unneeded complexity.

Another useful outcome of MD studies is the charge of any atom A along the trajectory, $Q_A(\varsigma)$, which can be combined into segments, revealing the charge flow between functional groups during the reaction. These charges are easily obtained by extracting data in US MD, for a current value of ζ at each point in the trajectory.

3.6 Periodic Boundary Conditions (PBC)

PBC can only be used with FMO-DFTB (excluding LC-DFTB). To use PBC, a periodic box should be defined. The "box" does not have to be rectangular. It is defined by providing 3 lattice vectors. A general space group symmetry is not supported. If there is any symmetry other than translations, the only way to deal with it is to replicate elementary cells using rotations and reflections, making a larger cell that can be described as P1 space group (this operation can be conveniently done using Mercury[16]).

It is only possible to compute the Γ-point, for which the wave vector is zero. Two kinds of gradients of the energy are calculated in PBC, with respect to nuclear coordinates and lattice vectors.

PBC should normally be used in three dimensions. Although technically possible, one and two-dimensional periodicity are not fully functional. The common practical way to describe it is to define a three-dimensional cell with very long lattice vectors in the dimensions, for which no periodicity exists. This approach is used for surface adsorption (Section 6.6), where two lattice vectors describe the two-dimensional periodicity of the surface and the third vector is chosen to be suitably long, which is equivalent to having a large layer of vacuum in that direction with no atoms in the empty space.

With PBC, one can optimize the geometry of atoms only, lattice vectors only, or both simultaneously. An optimization of lattice vectors is done by setting LATOPT in $DFTB for RUNTYP = OPTFMO (e.g., LATOPT = 1). The only implemented constraint of the lattice geometry is to keep rectangular lattices orthogonal with LATOPT = 2. For all other cases, the angles between lattice vectors can change during optimization.

An optimization of lattice vectors is conceptually similar to an optimization in Cartesian coordinates, whereas an optimization of lattice parameters is like an optimization in internal coordinates. In FMO, only lattice vectors can be optimized. The values of the three lengths a, b, and c and three angles α, β, and γ (the lattice parameters) are printed for reference.

One troublesome point of using PBC is the "spilling". This refers to atoms leaving the primary cell during geometry optimizations and MD. With PBC, it does not matter from the physical viewpoint whether the Cartesian coordinates of atoms are all in one cell, or if some atoms wandered off to a different cell. However, it affects computational efficiency. The algorithm of doing Ewald summations becomes extremely slow when a spilling extends into far cells.

To deal with a spilling, it is possible to push atoms into the primary cell, known as the wrapping. This is done by default for all atoms, and the wrapping can be stopped by adding 16384 to MODPRP in $FMOPRP. The wrapping by default is applied atomwise; there is also a fragment-wise wrapping, accomplished by adding 32768 to MODPRP. Any wrapping affects coordinates inside the FMO driver, whereas external engines (geometry optimizer or MD) use unwrapped coordinates. These options are summarized in Table 39.

To wrap atoms for good, the wrapped coordinates can be printed in a preliminary run by setting the output level to 0 in NPRINT of $FMOPRP (for example, using NPRINT = 8), then these wrapped coordinates can be copied to a new input as $FMOXYZ.

There is an option NSPILL in $DFTB to control how far the atoms are spilled (how many cells away from the original cell, in any of the

Table 39.   Suboptions in MODPRP related to atom wrapping in PBC.

Value[a]	Meaning
0	Wrap spilled atoms individually.[b]
16384	Do not wrap any spilled atom.
32768	Wrap spilled atoms as whole fragments.

[a]The indicated value should be added to the packed option MODPRP.
[b]A visualization of a wrapped molecule may be shocking to someone not used to PBC (a Picasso touch).

three directions). In FMO, NSPILL has to be given if the initial geometry has spilled atoms, but if any further spilling occurs during a geometry optimization or MD, then the spilling parameter NSPILL is internally adjusted. By default, the wrapping is done so no spilling can occur.

Another option that sometimes requires a manual definition in PBC is MAXINT in $DFTB. It refers to the buffer size allocated for an array storing a list of atoms within a certain radius from each atom. For a high density of interacting atoms, MAXINT has to be increased from its default value of 200.

3.7 Geometry Optimizations

There are two main engines to do an FMO geometry optimization, denoted by RUNTYP, OPTIMIZE and OPTFMO. There is also a special kind of a constrained optimization, SADPOINT, to locate transition states (or saddle points, which are not sad, although their search can be), and a search for a minimum energy crossing of two electronic states (MEX). The restrictions and features of SADPOINT are similar to OPTIMIZE. A summary is given in Table 40.

A geometry optimization proceeds until it either hits the limit of the number of steps NSTEP or the gradient becomes smaller than a threshold OPTTOL = ε (namely, the maximum gradient element is less than ε and the gradient RMS is less than $\varepsilon/3$).

Delocalized coordinates (DLC) are internal coordinates automatically generated from the Cartesian coordinates in $FMOXYZ. They can be used in RUNTYP = OPTIMIZE by setting $ZMAT DLC = .T. AUTO = .T. $END and $CONTRL NZVAR = 1 $END. DLC is cumbersome to use if there are several separate molecules (for example, multiple water molecules), because pseudo-bonds between them have to be manually defined with NONVDW in $ZMAT in order to define internal coordinates for connecting molecules. But for 1 molecule (e.g., a protein), DLC works automatically. It is possible to freeze internal coordinates with DLC.

Table 40. Geometry optimization engines for FMO.

Property	OPTIMIZE (SADPOINT)	OPTFMO
Input group	$STATPT	$OPTFMO
Engine	general	native to FMO
Max atoms	2000[a]	none[b]
Transition state search	yes	no
Gradient-based methods	[yes][c]	yes
Hessian-based methods	yes	yes
Hessian diagonalization	required	never done
Hessian matrix[d]	replicated	distributed
Internal coordinates	yes (DLC)	no
Frozen domain	yes	no
Freezing Cartesians	yes	yes
Freezing internals[e]	yes (DLC)	no
Constraining internals[e]	yes	yes
Lattice optimization	no	yes

[a]MXATM.
[b]The current record is 1 million atoms.
[c]Some are available but seldom used.
[d]The Hessian matrix is allocated in full size on each core for OPTIMIZE, whereas the inverse Hessian is distributed among cores for OPTFMO.
[e]Freezing means absolute fixation; constraining means harmonic constraints with a user-provided force constant.

Sometimes an optimization results in energy oscillations. They may be caused by a gradient inaccuracy, but they may also be due to problems in the optimization engine itself. In the former case, consider using SCZV, increasing RESPPC, and applying gradient projection. For engine problems, there are options in RUNTYP = OPTIMIZE to reduce the step size, such as DRMAX (for instance, it can be reduced to 0.1).

A simple summary and advice to the choice of OPTIMIZE vs. OPTFMO is as follows. One step of OPTIMIZE is more costly (a Hessian matrix is diagonalized at each step), but the number of steps to convergence is usually smaller; OPTIMIZE is recommended when the number of atoms is roughly 1000 atoms or less. OPTIMIZE can get confused by a negative curvature of the Hessian (in this case, restart the optimization). OPTFMO seems to work better with PBC. As some features are available in one method only, such as transition state search, there may be no choice for those tasks. The maximum number of atoms for OPTIMIZE in FD applies only to the size of the polarizable domain **B** (see Section 4.7).

To fix bond lengths or angles (including dihedrals), the most convenient option is to add a penalty harmonic potential in Eq. (61) for every desired constraint of an internal coordinate. This is done by defining internal coordinates in IHMCON, for example, IHMCON(1) = 1,2,8, 2,5,6,7 defines two of them, a stretch (type 1) between atoms 2 and 8 and an angle (type 2) between atoms 5, 6, and 7. The values of ζ_0 can be defined as SHMCON(1) = 1.0, 120.0 (1 Å and 120°) and the force constants k as FHMCON(1) = 100,20. In optimization and MD, the units for k are kcal/mol divided by $Å^2$ or $degree^2$, depending on the nature of ζ. The larger the force constant, the closer the value of ζ will be to the desired ζ_0 in the relaxed structure.

A summary of options for RUNTYP = OPTFMO is given in Table 41. METHOD = HSSUPD is in general the best optimization method, which requires storing the inverse Hessian matrix, distributed among all cores (allocated in MWORDS). If memory is of concern, or an energy surface is flat, one can try the gradient-only method CG.

Table 41. Options in $OPTFMO.

Option	Meaning
METHOD = A	HSSUPD (Hessian update), CG (conjugated gradient), STEEP (steepest descent).
OPTTOL = X	Optimization threshold applied to gradient, in hartree/bohr (default X = 1e-4).
NSTEP = K	Maximum number of steps (K = 200).
IFREEZ	List of individual Cartesian coordinates to freeze, which are sequentially numbered starting from x of atom 1 and up to z of the last atom.

Table 41.   (*Continued*)

Option	Meaning
IACTAT	List of atoms to be optimized (the rest is frozen). IACTAT(1) = −1 can be used to freeze all atoms in lattice optimizations.
IHMCON	List of definitions of internal coordinates, composed of sets. Each set begins with a type: 1 (stretch), 2 (angle), or 3 (dihedral angle), followed by atoms. For example, 2,12,15,18 defines the angle between atoms 12, 15, and 18.
FHMCON	Force constants k for each coordinate, as in Eq. (61).
SHMCON	Coordinate values ζ_0 in Eq. (61), in Å for stretches, or degrees for angles.
STEP = A_0	Scaling factor to multiply steps in coordinates.
STPMIN = A_1	Minimum value of such factor.
STPMAX = A_2	Maximum value of such factor.
STPFAC = A_3	A constant for dynamically adjusting the factor.
NPRICO = K	Print coordinates every K steps.
IREST = 1	Write full restart data every step (seldom used).
NFGD = N	Number of fragments (same as NFRAG in $FMO), required for MODIO = 3072 in $SYSTEM or MODPAR = 1024 in $FMOPRP.
MAXNAT = K	Maximum number of atoms per fragment. MAXNAT can be specified optionally to reduce I/O in DFTB.

There are two ways to freeze Cartesian coordinates, either by specifying IFREEZ for coordinates to freeze (the rest is optimized) or IACTAT of atoms to optimize (the rest is frozen). IFREEZ operates on individual x, y, and z coordinates. For example, IFREEZ(1) = 5,9,10 freezes y coordinate of atom 2, z coordinate of atom 3 and x coordinate of atom 4 (all coordinates are considered as a list like $x_1, y_1, z_1, x_2, y_2, z_2, x_3, y_3, z_3$, etc. (where atoms are subscripts), represented by numbers 1,2,3,4,5,6,7,8,9, etc. In contrast, IACTAT operates on atoms, for example, IACTAT(1) = 5,9,10 optimizes atoms 5, 9, and 10.

If STPFAC is 1, then the scaling factor STEP (which multiplies the predicted coordinate step) is constant during geometry optimization. Otherwise, if energy goes down in an expected way, the step size is divided by STPFAC; if the energy goes up, STEP is multiplied by STPFAC. A typical value of STPFAC is 0.9, corresponding to a 10% change. The values of the dynamically adjusted STEP are limited to the range [STPMIN;STPMAX]. A typical value of STEP is 0.7, whereas STPMIN may be 0.3 and STPMAX 1.0.

NFGD is a redundant definition of NFRAG, needed for some acceleration options. If MAXNAT is defined, then the charge restart file used by DFTB can be replaced by an in-memory array. The value of MAXNAT may be taken from an output of the same job, "Array dimensions maxbnd,maxknd,maxcbs,maxcao,maxbbd,maxnat are:", the last number printed.

If a $PDB is given in the input for RUNTYP = OPTFMO, then the optimized structure is written to DAT in the PDB format, which may be convenient for optimizing biochemical structures. In the optimization, the initial coordinates in the $PDB are ignored, and the initial geometry is taken from $FMOXYZ.

3.8 Molecular Dynamics (MD)

FMO/MD can be done with any method that has analytic gradients. It is recommended to use the fully analytic gradient including SCZV responses (computed by MODGRD = 42 in $FMO, except that for DFTB with HOP and AFO the values are 32 and 48, respectively). Except for DFTB, RESPPC should either be set to 0 (turning off the approximation) or to a sufficiently large value (at least 2.0) due to the gradient accuracy issues (for DFTB, the default value of RESPPC = 2 is used). See Section 2.6.2.8 for more details on approximations.

Not all methods have an accurate analytic gradient, for example, FMO-MCSCF has only an approximate gradient. Some methods such as CC or TDDFT/PCM have no analytic gradients.

3.8.1 *Creating Input Files for MD*

In MD, it is recommended to reduce the output to the shortest level to make the output file smaller. The most convenient way is to set MODIO = 3072 in $SYSTEM and avoid setting NPRINT in $FMOPRP.

It is useful to change the trajectory file format. By default, GAMESS uses its own native trajectory format that only MacMolPlt can read. For a widely supported multiple XYZ format, use TRJFMT = F14o6 in $MD. Here, 14 is the number of decimal places per coordinate, and 6 is the number of digits after the decimal point (like the format of F14.6 in FORTRAN, but because GAMESS cannot process dots in the text values, a letter "o" is used). A trajectory file can be made a little smaller by choosing a suitable format in TRJFMT.

The MD engine in GAMESS has two ensembles, NVE and NVT, but NPT is not available. Some useful options related to MD are shown in Table 42. With RATTLE, the recommended time step is 1 fs (DT = 1e-15). There are no limits to the duration and the number of atoms. Currently there is no way of computing RDF automatically, but external tools can be used (Section 3.8.2).

JEVERY defines how often MD properties such as the temperature or kinetic energy are printed and how often energies are averaged in

Table 42. Useful options in $MD.

Meaning	Options	Comments
Time step	DT = 1.0D-15	1 fs.
Number of steps	NSTEPS = 1000	Total duration DT × NSTEPS = 1 ps.
NVE	NVTNH = 0	NVE is mainly used for testing.
NVT	NVTNH = 2	NVT.
	RSTEMP = .T.	RSTEMP, MBT, and MBR are here set for
	MBT = .TRUE.	starting anew; for restarts, use the values
	MBR = .TRUE.	from a restart file.
	BATHT = 100	Temperature 100 K.
RATTLE	IRATTL(1) = 1,1,1	Stabilize larger time steps.
Fluctuation analysis	MODFLU = 1	Average energies along the trajectory.
Umbrella sampling	USAMP = .T.	Use US,
	IUSTYP = 0	ζ is a bond distance (type 0),
	RZERO(1) = 1.2	$\zeta_0 = 1.2$ Å,
	IPAIR(1) = 3,6	$\zeta = |\mathbf{R}_3-\mathbf{R}_6|$, the two atoms are 3 and 6,
	UFORCE = 20	k in Eq. (61), in kcal mol^{-1} Å^{-2}.

(Continued)

Table 42. (*Continued*)

Meaning	Options	Comments
Spherical boundary	SSBP = .T. SFORCE = 2.0 DROFF = R	Atoms in a spherical droplet, constrained by a harmonic potential with $k = 2.0$ for radius R.
Centering potential	CCMS = .T. CFORCE = 1.0 NATCC = K	Solute origin-centering potential with $k = 1.0$, applied to solute atoms 1…K.

FA (for averaging at every step, use JEVERY = 1). On the other hand, KEVERY determines how often the restart and trajectory files are updated. Making these values smaller increases the output and trajectory file sizes. MD restarts are possible, described in Section 5.5.2.

Among other useful things, there is a harmonic potential for a spherical boundary SSBP, which is used to prevent "evaporation" from a spherical droplet, when PBC are not used. The centering potential CCMS is sometimes used with such a spherical droplet to prevent the solute from sticking to the "wall" and making sure it stays in the center, properly solvated in the droplet. For CCMS, all solute atoms should be in the beginning of the atom list in $FMOXYZ.

3.8.2 *Processing Results of MD (RDF)*

To compute free energy for constrained MD, the values of an US coordinate can be extracted from an output file, and WHAM[22] can be used to process the results (Section 3.10.2).

For calculating radial distribution functions (RDF), VMD can be used. A trajectory should be generated with TRJFMT, and renamed to have an xyz extension (it is a multiple structure XYZ file). Then this trajectory can be read into VMD and "gofrgui" plugin can be executed from the command prompt in VMD. Some periodic box should usually be defined in the Utilities menu of gofrgui. Computing RDF requires specifying two selections. For example, "NAME Ca" and "NAME O". This will calculate the RDF for Ca–O distances. The coordination number of Ca in water can be obtained by integrating the first RDF peak.

3.9 Hessians, IR and Raman Spectra

A Hessian is a matrix of second derivatives of the energy with respect to nuclear coordinates, scaled by square roots of atomic masses. Hessians are very useful to find transition states of chemical reactions, calculate vibrational frequencies, thermodynamic properties, and simulate IR and Raman spectra.

Analytic second derivatives are time and memory consuming, and they are difficult to derive and implement. Alternatively, one can do a 2-point numerical differentiation of analytic gradients known as semi-numerical derivatives, which on the one hand is not memory consuming and readily possible whenever there is a good quality gradient, but on the other hand one has to do many single point gradient calculations. If it is not specified in the input file whether the Hessian should be computed analytically or semi-numerically, then the former is chosen if available.

There are two different Hessian engines for FMO selected by RUNTYP, HESSIAN and FMOHESS (Table 43). The main engine is

Table 43.   Hessian engines for FMO.

Property	HESSIAN	FMOHESS
Input group	$FORCE	$FORCE
Max atoms	2000[a]	none[b]
Analytic	No	Yes
Semi-numerical	Yes	Yes
with restarts	Yes	No
Purification[c]	Yes	No
Isotopes	No	Yes
Frozen domain	Yes	Yes
Partial scheme[d]	PHA	PAVE
Thermochemistry	Yes	Yes
IR intensities	Yes	Yes
Raman activities	Yes	Yes

[a]MXATM.
[b]Thermochemistry and spectra for analytic Hessians only are limited by MXATM.
[c]Transformation of the Hessian to make rotational and translational frequencies zero.
[d]Partial Hessian for a subset of atoms.

FMOHESS. The maximum number of atoms for frozen domain (see Section 4.7) applies only to the size of the polarizable domain **B**. Isotope effects on thermodynamic quantities and spectra can be taken into account by defining atomic masses of all atoms as AMASS in $MASS. The default values for most abundant isotopes are predefined, so one can change only the masses of some atoms, for example, AMASS(346) = A for changing the mass of atom 346 to be A (au).

A semi-numerical Hessian is chosen with METHOD = SEMINUM in $FORCE (possible with both RUNTYPs), whereas analytic Hessians can be done with METHOD = ANALYTIC only for RUNTYP = FMOHESS.

As a current peculiarity, IR intensities for RUNTYP = HESSIAN with MP2 can only be computed with CODE = IMS (for other MP2 codes the intensities come out as 0).

The purification of the Hessian (PURIFY = .T. in $FORCE) is a transformation resulting in 3 translational and 3 rotational eigenvalues to be zero for non-linear molecules. Without the transformation, due to various errors (DFT grid, PCM discretization, gradient being not zero, approximations in the gradient and/or Hessian, etc.), some eigenvalues that should be zero are not.

In a partial Hessian, semi-numerical second derivatives are computed for a subset of atoms, for example, for atoms in a binding pocket of a protein. RUNTYP = HESSIAN uses partial Hessian analysis (PHA), whereas RUNTYP = FMOHESS employs PAVE (Section 3.5.7). In the former, the Hessian matrix has full dimensions, no matter how small the subset of atom is; the partial block for active atoms is calculated and the rest is filled with zeroes and small numbers on the diagonal. The purpose is to purify the Hessian, which is only possible when all atoms are included. In contrast, the other engine PAVE constructs only the block for selected atoms.

A useful option for PAVE is NPRHSS = 16 in $FORCE, which punches the full-size Hessian (the PAVE block with appended zeros for other blocks). This Hessian can be used for transition state search in the full coordinate space.

3.9.1 *Vibrational Entropy and Enthalpy*

To calculate thermochemistry (vibrational enthalpy, entropy, and free energy), it is recommended to optimize geometry to a tighter threshold, OPTTOL = 1e-5 in $STATPT or $OPTFMO. This is especially true for the entropy and free energy, which are determined by low frequency vibrations. The ZPE, on the other hand, is more forgiving as it is determined mainly by large frequencies.

Thermodynamic quantities are calculated by default in a Hessian run. The vibrational frequencies are square roots of the Hessian eigenvalues, so any negative eigenvalue results in an imaginary frequency. The list of normal modes is arranged as follows. First, imaginary eigenvalues are listed, if any, then real values. In the calculation of thermochemistry, imaginary and, optionally, some real frequencies are skipped (RUNTYP = HESSIAN makes a distinction between large and small imaginary frequencies; the latter are treated as numerical noise on par with small real values).

Vibrational energies with the exception of ZPE are temperature dependent. It is possible to calculate them for up to 10 temperatures in TEMP of $FORCE. Frequencies can be scaled with a global factor, specified in SCLFAC in $FORCE (in an attempt to correct for systematic errors, for example, due to anharmonicity).

RUNTYP = FMOHESS calculations cannot be restarted, even semi-numerical ones. For RUNTYP = HESSIAN, restarts are possible, as for non-FMO runs, by manually pasting restart data from the DAT file of a preceding calculation into the input file.

For semi-numerical derivatives the coordinate shift VIBSIZ(1) in $FORCE is important. Its value (in bohr) can have a large effect on vibrational entropies and free energies, because it affects low frequencies. On the other hand, if one is interested in ZPE or a transition state search, the value of VIBSIZ(1) is less important. Using a large value (> 0.01) is a bad idea because mathematically the differencing can produce large errors. Using a small value (< 0.001) is often a bad idea as well, because for this case, numerical errors in the calculations can affect the results. Therefore, in most cases, the values of 0.005–0.01 are recommended.

In terms of the accuracy of analytic Hessians, only for DFTB in vacuum the exact second derivatives can be obtained; for other methods there are some approximations, which are somewhat larger for DFT and PCM (semi-numerical Hessians may have a higher accuracy than approximate analytic Hessians). Analytic Hessians are not available for EFP, MCP, MCSCF, TDDFT, MP2, etc.

3.9.2 *IR Spectra*

Peaks in IR spectra are found at vibrational frequencies, and the intensities are obtained from the derivatives of dipole moments with respect to nuclear coordinates. For analytic Hessians, these derivatives are calculated analytically, and for semi-numerical Hessian, numerically. IR intensities are computed by default in a Hessian calculation.

IR spectra thus obtained are discreet, because they lack broadening due to temperature (rovibrational effects). The only way to take into account these effects is via a Gaussian broadening function, which can be done by means external to GAMESS.

3.9.3 *Raman Spectra*

Peaks in normal Raman spectra are also at regular vibrational frequencies, and Raman activities require second derivatives of the energy gradient with respect to the static electric field. These derivatives are always calculated numerically. The number of these single point calculations is 19 (1 + 18, as given by the number of unique combinations of Cartesian indices), independent of the number of atoms.

Raman spectra can be simulated in two ways. One is to calculate a full Hessian, take it as $HESS from the DAT file, paste it into a new input and perform a RUNTYP = RAMAN calculation, which does 19 single point gradient calculations. The other way is to specify VIBSIZ(2) in a semi-numerical RUNTYP = FMOHESS calculation, in which case Hessian, IR, and Raman spectra are obtained in a single run. The recommended value of the field derivative is 0.002, specified as VIBSIZ(2) in $FORCE for RUNTYP = FMOHESS or EFIELD in $RAMAN for RUNTYP = RAMAN.

Similar to IR, Raman spectra are discrete and a Gaussian function can be used to simulate the broadening effects.

3.10 Chemical Reactions

For treating chemical reactions in FMO, it is typical to include all reactants into one fragment to increase the accuracy. It is straightforward to treat a catalyst too, and there are examples using FMO to describe catalysis, for example, solid-state catalysis (Section 6.6) or enzymatic reactions.

There are two main routes, one via mapping of the reaction path with a structure optimization and another via molecular dynamics. In MD, temperature is inherently used, and for a single QM path, statistical thermodynamics can be used to get vibrational contributions at a given temperature. Temperature is also explicitly used in the computation of the rate constant from a reaction barrier.

3.10.1 Static Reaction Path (IRC)

A static path is obtained by finding an energy change $E(\zeta)$ from reactants to products via a transition state (TS), where ζ is a reaction coordinate. A TS is a maximum in the energy with respect to ζ and a minimum in all other coordinates. The only way to conduct a transition state search in this "classical" textbook way is to use RUNTYP = SADPOINT. This type of study is usually done by first guessing a transition state structure with semi-broken bond(s), stabilized by the environment, running a Hessian, confirming that there is a hopefully large imaginary frequency with a normal mode corresponding to the desired reaction, and then running SADPOINT, providing a Hessian in the input as $HESS.

There are two ways to accelerate these calculations. Approach A: use a partial Hessian expanded in the full dimension. For example, a reaction may involve 10 atoms but there are 300 atoms in the system.

It is possible to calculate a semi-numerical partial Hessian with RUNTYP = FMOHESS (a 30 × 30 matrix) and use the option NPRHSS = 16 in $FORCE to expand the partial Hessian into the full size (900 × 900). This full-size Hessian can be read in RUNTYP = SADPOINT and used for a transition state search.

Approach B: an alternative to this is to use frozen domain, and similarly, to compute an analytic or semi-numerical Hessian, and then run SADPOINT. In frozen domain, only a part of the molecular system is optimized.

Once a transition state is located, intrinsic reaction coordinate (IRC) can be used with RUNTYP = IRC to map the reaction path. It is done in two directions by setting FORWRD in $IRC to .T. or .F., one of which will lead to reactants and the other to products (which can be identified by looking at the two final geometries). IRC tends to "lose its way" in the rough energy surface landscape eventually, that is, it stops before reaching the minima for reactants and products. To complete the path, a geometry optimization can be done in each direction, starting from the final point found by IRC.

An IRC calculation needs a Hessian for a transition state (specified in $HESS) and its geometry. FORWRD should be .T. in one run, and .F. in another. Among other options, NPOINT in $IRC should be set to the desired number of points on the path (such as 20).

3.10.2 Dynamic Reaction Mapping (MD)

The dynamic approach is to use molecular dynamics. Although one could try to simply start MD at reactants and hope that the reaction might occur, it may take aeons. Currently, the most convenient way is to run a biased MD using umbrella sampling (US).

In MD/US, a reaction coordinate should be defined by the user *a priori*. It can be a bond length, angle, or dihedral angle, and some more complex choices are available (see docs-input.txt). Consider an S_N2 reaction in explicit water (see Section 6.6).

$$I\text{-}CH_3 + NH_3 \rightarrow I^- + H_3C\text{-}NH_3^+ \tag{70}$$

This reaction occurs starting at reactants (I–C...N) via a transition state (I...C...N) to products (I...C–N). The reaction coordinate (an asymmetric stretch) is conveniently chosen as the difference between I,C and C,N distances.

$$\zeta = R_{IC} - R_{CN} \tag{71}$$

For reactants, R_{IC} is small and R_{CN} is large, i.e., $\zeta < 0$, and for products it is the opposite ($\zeta > 0$). The transition state is not found at 0 because of the asymmetry (C moves between different elements, I and N). The reaction path thus should start at a negative value of ζ, pass through 0 and go to some positive value. When values of $|\zeta|$ are large, the reactants or products are far away, so there is no need to sample that region where nothing happens. One can choose the ζ range of −2 to 2 Å, at least, as an initial guess.

Next, one has to select grid points in the range. The more points, the more expensive the calculations, but taking too few points is also bad because the quality of the reaction path may suffer (a free energy profile may have discontinuities and kinks). One can begin with an equidistant set, for example, with a step of 0.4 Å, yielding the following points $\zeta_0 = -2, -1.6, -1.2, -0.8, -0.4, 0, 0.4, 0.8, 1.2, 1.6,$ and 2.0. For it, it is necessary to perform 11 MD simulations (sometimes called windows), and in each one to constrain the coordinate to be in the vicinity of the value of ζ_0 by applying a penalty potential in Eq. (61).

The force constant k may be chosen differently for each ζ_0. If the force constant is too large, one will have to take too many values of ζ_0, because the penalty is so high that the reaction coordinate may not sample the region between two windows. If it is too small, the simulation may not sample the parts of the reaction path with high energy (the transition state), because the molecular system will prefer to descend to a lower energy region as much as the constraint leash allows. Force constants of 10–100 kcal mol^{-1} Å^{-2} are often used.

It is common to relax the structure in MD, prior to doing a production run, which is known as a pre-equilibration. This can be done as two separate runs (a fresh MD run for pre-equilibration, followed by a restart MD for production, see Section 5.5.1), or one run of a combined total duration, in which the initial part of the trajectory is not used for data processing. The latter way is more convenient. A sample of the header of the MD group for the latter approach is shown in Table 44. As for the intimate question for how long to do pre-equilibration and production, it should be decided based on the available resources. The longer the better.

There is no general automatic solution to the choice of ζ_0 and k. After making a guess, and running MD, the quality of sampling should be checked. This can be done by extracting the values of ζ (Section 6.4) at every time step in the output files of all windows, plotting the values and checking if there may be any empty under-sampled portions, where too few points appear. In this case, an extra MD simulation can be performed to fill the gap.

Table 44.   Sample $MD for umbrella sampling.

Input	Comments
⌴$MD	
NVTNH = 2	NVT,
RSTEMP = .T. MBT = .TRUE. MBR = .TRUE.	from scratch (not a restart).
IRATTL(1) = 1,1,1 DT = 1.0D-15 NSTEPS = 200000	RATTLE, 1 fs step, 200 ps total.
BATHT = 303.0	T = 303 K.
JEVERY = 10 KEVERY = 1000	MD properties, restart and trajectory.
TRJFMT = F14o6	XYZ format for trajectory file.
MODFLU = 1	FA is enabled.
USAMP = .T. IUSTYP = 1	US potential, asymmetric stretch.[a]
UFORCE(1) = 50 IPAIR(1) = 1,6,5 RZERO(1) = −1.8	k = 50, $\zeta = R_{6,1}$-$R_{5,1}$, $\zeta_0 = -1.8$ Å.
⌴$END	

[a]$R_{A,B}$ is the distance between atoms A and B. In this system, I, C, and N atoms in Eq. (71) happen to be 1, 5, and 6, respectively.

Table 45. **Example of extracted reaction coordinates $\zeta(t)$ from an US MD for RZERO(1) = -2.0 simulation in the preparation for computing PMF, saved as file "usmd.z = -2.0.coord".[a]**

0	−1.9793301596
1	−1.9758524803
2	−1.9718905695
3	−1.9672757844
4	−1.9618906024

[a]The first column (t) in the file is an index that does not affect the results.

To calculate a free energy estimate, known as the potential of mean force (PMF), a file per US window has to be created (Table 45), with a pair of values of time step t and $\zeta(t)$ per line. For the above example, 11 such files can be created.

Table 46. **Example of a header file usmd.wham for running WHAM to get a PMF, listing trajectory file names, ζ_0, and k.[a]**

usmd.z = −2.0.coord	−2.0	20
usmd.z = −1.6.coord	−1.6	20
usmd.z = −1.2.coord	−1.2	20
usmd.z = −0.8.coord	−0.8	50
usmd.z = −0.4.coord	−0.4	50
usmd.z = 0.0.coord	0.0	20
usmd.z = 0.4.coord	0.4	20
usmd.z = 0.8.coord	0.8	30
usmd.z = 1.2.coord	1.2	30
usmd.z = 1.6.coord	1.6	20
usmd.z = 2.0.coord	2.0	20

[a]To execute WHAM, the following options can be used: "wham −2.0 2.0 200 0.0001 0 usmd. wham usmd.res".

Then a WHAM[22] input file can be created specifying a list of file names containing the coordinates, and values of ζ_0 and k used for each of them. The exact options for running WHAM have to be adjusted (angles may have periodicity). An example is shown in

Table 46. Executing WHAM, one obtains the PMF, which is a measure of the free energy, as a function of the reaction coordinate.

3.10.3 *Crossing of Energy Surfaces (MEX)*

A crossing of two energy surfaces for electronic states with a different spin can be used to estimate the barrier for a spin-forbidden process. This is done with a minimum energy crossing (MEX) search using two FMO calculations for each spin state and combining their gradients in a constrained optimization (the constraint is that the energy of the two states be the same). This type of calculation, selected with RUNTYP = MEX in $CONTRL, can be used for modeling catalysis involving transition metals. A MEX calculation is sometimes called a minimum energy crossing point (MECP) search. MEX is different from a conical intersection search (for surfaces of the same spin), with which it should not be confused.

A MEX calculation is a constrained optimization of 2 spin states. For each step, two calculations are performed, one for each spin multiplicity, for example, for a quartet and a sextet (as is often the case for iron-containing compounds). Then a next point in the optimization is estimated using two gradients, and the energy minimization of both states proceeds under the constraint that their energies should be equal.

MEX presents a certain challenge for FMO due to having to keep data for two different spin states at the same time. The only way to handle it is to use the three-level GDDI (Section 5.10.3). This can be thought of as two independent regular two-level GDDI calculations running concurrently, and exchanging data as needed. Each world in GDDI/3 (for terminology, see Section 5.10.3) stores a separate F40 containing the data for its spin state. MEX should not be used with DDI or regular two-level GDDI.

For MEX, NGROUP in $GDDI should be 2 (two worlds), and NSUBGR should define the number of groups, the same for each multiplicity. For example, if 8 nodes are used, they can be divided into 2 worlds (NGROUP = 2) of 4 groups each (NSUBGR = 4).

In FMO, the two spin states should use the same options except for multiplicity and SCF type. Typically, the SCF choices in MEX are

Table 47.   Input keywords for MEX (state 1: RHF singlet, state 2: ROHF triplet).

Group	Meaning
⌴$CONTRL RUNTYP = MEX	MEX
SCFTYP = ROHF $END	SCFTYP is set to an open-shell type.
⌴$GDDI NGROUP = 2 NSUBGR = M $END	2 worlds in GDDI/3, M groups per world.[a]
⌴$MEX IMEXFG = 4	Fragment 4 is the active fragment.
MULT2 = 3	Multiplicity of state 2.
SCF2 = ROHF $END	SCF of state 2.
⌴$FMO NFRAG = 4 SCFTYP(1) = RHF	SCF type for state 1.
SCFFRG(1) = RHF,RHF,RHF,RHF	SCF types for each fragment (state 1).
MULT(1) = 1,1,1,1 $END	Multiplicity of each fragment (state 1).

[a]M can be a suitable positive integer, such as 1. NGROUP should be 2.

RHF, ROHF, and UHF (also, RDFT and UDFT). MEX can be used when only one active fragment, selected in IMEXFG of $MEX, changes its spin state. The first spin state should be set up in $FMO (as SCFFRG, SCFTYP, and MULT). For the other spin state, MULT2 in $MEX defines its multiplicity and SCF2 selects its SCF type. A sample header is shown in Table 47. In the example, the active fragment is the last, number 4, with three solvent fragments 1, 2, and 3. Two spin states are computed, a singlet (RHF) and a triplet (ROHF).

3.11 Atomic Charges, Multipole Moments, and Spin Populations

Atomic charges are calculated for the RESPPC approximation (Section 5.2) with three choices for atomic charge schemes (Mulliken, Stone, and ESP). To print these charges, add 8 to NPRINT in $FMOPRP, for example, use NPRINT = 9. In DFTB, only Mulliken charges can be used.

In FMOn calculations, m-body atomic charges are computed for $m \leq n$ (m can be specified as MODCHA in $FMO). One-body charges reflect polarization, whereas higher-order charges also incorporate CT. Some acceleration can be achieved in FMO3 by preventing the calculation of FMO3 CTs by setting MODCHA = 2.

By default, Mulliken charges are used, which is the only fully supported type. The other types of charges have limitations to their usage, for example, there are no analytic gradients for them. The chosen charge type is used for the RESPPC approximation in the embedding potential, so it affects the electronic state.

The basic options for selecting Stone and ESP charges are shown in Table 48. Both kinds of charges have various detailed options that can be modified if necessary (for example, it is possible to use numerical integration for diffuse functions by specifying a value of BIGEXP in $STONE). ESP charges are fitted to reproduce the ESP on a set of grid points with an optional constraint of preserving the dipole (i.e., the dipole from charges should match the dipole from the electron density).

Multipole moments in Eq. (15) can be computed up to octupoles (in DFTB, only dipole moments can be obtained). The maximum order of the multipoles is set as IEMOM in $ELMOM, for example, IEMOM = 2 will print quadrupole moments. The origin of multipoles is selected with WHERE in $ELMOM (the default choice is COMASS, center of mass). In FMO the same origin is used for all n-mers.

Dipole moments for neutral systems do not depend on the origin R_0, but in general multipole moments depend on it. In PCM, the solvent contribution to dipoles is computed, which is a solvent screening for dipoles. The total solute + solvent dipole moment shows a weak dependence on the origin, because the sum of the solute

Table 48.   Options for Stone and ESP charges.

Type	Options
Stone	$STONE
	ATOMS
	$END
ESP	$ELPOT IEPOT = 1 WHERE = PDC $END
	$PDC PTSEL = GEODESIC CONSTR = DIPOLE $END

charge Q plus the induced solvent charge Q^{solv} is $Q + Q^{solv} \approx Q/\varepsilon$, a small value for polar solvents.

Solvent dipoles are computed using the induced solvent charge q_i and coordinates $\mathbf{R}_i$ of tessera i.

$$\mathbf{d}^{solv} = \sum_{i=1}^{N^{TS}} q_i (\mathbf{R}_i \text{-} \mathbf{R}_0) \tag{72}$$

The charges are always computed with the same density as used in the embedding (in MP2, RHF density is used to get the charges). The multipole moments and properties on a grid, however, are computed with the highest-level density available, which means that for MP2 gradients (or if MP2PRP in \$MP2 is set to .T.), MP2 density is used.

In UHF or UDFT calculations there is a spin density for a fragment X, which is defined as the difference of the electron densities for α and β spins. By convention, all unpaired electrons have α spin, for example, there are two unpaired α electrons in a triplet.

$$\mathbf{D}^{X,spin} = \mathbf{D}^{X,\,\alpha} - \mathbf{D}^{X,\beta} \tag{73}$$

whereas the electron density is their sum,

$$\mathbf{D}^X = \mathbf{D}^{X,\,\alpha} + \mathbf{D}^{X,\beta} \tag{74}$$

The spin density shows where the radical character is localized. It is possible to plot the total spin density of the whole system or some fragment (Section 4.9) and compute Mulliken atomic spin populations (analogues of atomic charges), which are useful to analyze where the spin density is localized.

Likewise, spin transfer $\Delta S_{I \to J}^{spin}$ can be calculated, an analogue of CT. $\Delta S_{I \to J}^{spin}$ is computed by subtracting atomic spins S_A^{spin} in dimer IJ and two monomers I and J for atoms A in IJ. This quantity shows how much spin density is transferred from one fragment to another, a possibly useful quantity for spintronics.

CHAPTER 4

Building up Complexity

4.1 Parametrized Methods

4.1.1 *Density Functional Theory (DFT)*

In FMO-DFT, all fragments use the same functional within each layer, but it is possible to define two layers with different functionals. One can also define an RHF layer and a DFT layer, which may be a good compromise between the SCF instability of DFT and the benefits it offers in treating the electron correlation. DFT can be restricted (RDFT) and unrestricted (UDFT), but not RODFT. Multiple open-shell fragments are allowed, except for analytic second derivatives.

With respect to FMO, the main problem of DFT is that some functionals tend to produce a small gap between occupied and virtual orbitals. This has two consequences: a bad convergence and a poor accuracy of fragmentation, so that to get a decent accuracy one has to use larger fragments and often a higher order MBE than for HF.

Two factors have a large effect on the gap: the fraction of the HF exchange (the larger the better) and the use of a long-range correction (LC) in LC-DFT. Among the LC-DFT functionals implemented in GAMESS, CAM-B3LYP is one of the most stable for FMO. A polar solvent tends to increase the gap by preferentially stabilizing occupied orbitals more than virtuals, so that PCM usually stabilizes DFT (and HF) calculations.

To deal with the convergence problem, there are some FMO-specific tricks. Each fragment calculation is computed multiple times in the monomer loop to equilibrate the embedding. One trick is to turn off DFT (use HF) for a few initial iterations in the monomer loop.

This is done by defining COROFF in $FMOPRP. For example, COROFF = 1e-3 (which is the default) will run a few iterations with HF until the monomer loop convergence reaches this value (usually in 4–6 iterations) and then DFT is turned on. When SWOFF is set in $DFT (e.g., SWOFF = 1e-3), DFT is switched off in the initial SCF iterations of dimers and trimers (but not monomers).

DFT convergence can be excruciating and may call for a dexterous use of the full panoply of SCF options (Section 4.10). Open shells and quasidegeneracies can make bad things worse, and in challenging cases a fragment may take many dozens of struggling iterations trotting toward convergence (degrading GDDI performance if more than 1 group is used, see Section 2.9.2).

As far as DFT grids are concerned, there is nothing special to FMO (except that the grid quality switching option is not used in FMO). The default is a Lebedev grid and there is a fast but less accurate SG1 grid. The use of a grid in DFT tends to destroy the rotational invariance of the gradient, so at the end of an FMO-DFT calculation, the gradient is projected to preserve the property, which, however, is alike cutting a wound off rather than healing it. The projection may negatively affect the gradient accuracy, yet it is usually better to project the gradient than not. The SG1 grid can produce some numerical noise in the gradient, which can be vexing in geometry optimizations.

4.1.2 *Corrections for SCF (DFT-D and HF-3c)*

In HF with 3 corrections (HF-3c), there are three force field-like corrections added to the HF energy, of which one is for dispersion (D3(BJ)) and the other two are related to reducing BSSE: geometric counterpoise correction (GCP) and short-range basis set incompleteness (SRB).

The basis sets for HF-3c are predefined and they should be used as provided (for most chemical elements they are of the minimum quality, but for some elements polarization functions are included). HF-3c can be used by setting GBASIS = HF-3c in $BASIS and defining hybrid orbitals in $FMOHYB (Section 4.11.1). If desired, individual corrections can be turned off, but it is seldom done. No parameters are predefined for DFT-3c.

One of the three models in HF-3c, the dispersion correction (DC), can be used in a standalone way with HF, DFT, and DFTB, denoted as DFT-D for DFT. The parameters for the dispersion are internally set depending on the basis set and the functional. If a basis set is entered manually, then the user should also manually define all dispersion-related parameters in $DFT.

D2, D3, and D3(BJ) dispersion models can be chosen by selecting $DFT DC = .T. IDCVER = 2, 3, and 4, respectively (the $DFT group can be used to define dispersion in HF and DFTB too).

D3 and D3(BJ) models have a small approximation in the FMO gradient unless the bulk dispersion is chosen with MODPRP = 4096 (the bulk option has a restriction of MXATM atoms for the total system). Except for DFTB, dispersion energy can be added for ES dimers by MODPRP = 8192 in $FMOPRP (for UFF in DFTB, it is always added). In DFTB/PBC, only UFF dispersion may be used.

4.1.3 *Density-functional Tight-binding (DFTB)*

DFTB is a fast parametrized QM method, based on an expansion of the electron density in a Taylor series from the reference point of neutral atoms. Different DFTB levels correspond to a truncation of the series to 0-order in non-charge-consistent (NCC) DFTB, up to second order in self-consistent charge (SCC) DFTB (also called DFTB2), or up to third order (DFTB3). First-order DFTB is not used, because due to charge conservation it is equivalent to 0-order.

Traditionally, DFT is the underlying theory with which DFTB parameters are derived, although one term (the repulsion energy) remains for an arbitrary fitting, to some theory and/or experiment. There is a long-range corrected DFTB2 (LC-DFTB2), derived from LC-DFT.

In DFTB, only valence electrons are retained (as in core potentials). Because there are no core electrons, DFTB has a reduced BSSE, and the interaction energies in DFTB are close in magnitude to the values in regular QM methods for large basis sets.

For alkali cations, there are no electrons, because the only valence electron is removed. Such electron-less fragments may be defined for computing interactions of ions with other fragments.

Integrals in DFTB are stored on a grid as parameters so their use is very fast, and AOs are not directly used in DFTB calculations. DFTB parameters are stored in a file per pair of elements, such as C–H. No atomic types are used, so all carbon atoms use the same parameters, and their hybridization is handled by the wave function. AOs in DFTB are used only for some special purposes such as computing density on a grid.

DFTB is a QM method, with a Hamiltonian, a wave function, and an electron density (corresponding to a closed-shell singlet state). Non-polarizable NCC-DFTB is seldom used, whereas DFTB2 and DFTB3 use a polarizable point charge embedding. The embedding is updated in an SCF-like process, called SCC in DFTB, so that a DFTB calculation has iterations and looks very much like an HF or DFT calculation.

In addition to parameter files, there are other options that can be specified in the input, such as Hubbard derivatives, details of damping functions, etc. The Dm and UFF models of dispersion, PBC, PCM, and SMD may be used with DFTB. There are analytic first and second derivatives, IR and Raman intensities.

For choosing a parameter set in DFTB, one should check if all desired chemical element pairs are included in it. LC-DFTB parameters are generated for a specific value of the range-separating parameter ω. Parameters are distributed from the DFTB homepage, and their distribution is subject to certain rules (that is, a license).[23] In general, parameters in GAMESS are compatible with those in DFTB+,[23] but not with other programs. Three sets of parameters are distributed with GAMESS, shown in Table 49. For each parameter set, its prescribed DFTB method should be used. NDFTB = 2 is used for DFTB2, whereas DAMPXH = .T. and NDFTB = 3 can be set for DFTB3. For LC-DFTB, LCDFTB = .T. should be also defined and EMU should be set to the value of ω, for which parameters were generated (see their readme file). Alternatively to NDFTB = 2, SCC-DFTB can also be turned on with SCC = .T.. UFF dispersion can be added with DISP = UFF.

Table 49. Input for DFTB parameter sets shipped with GAMESS.

Name	System	Method[a]	Input[b]
3ob-3-1	bio	DFTB3/D3(BJ)	$DFTB NDFTB = 3 DAMPXH = .T. PARAM = 3OB-3-1 $END $DFT DC = .T. IDCVER = 4 $END
matsci-0-2	inorganic	DFTB2/UFF	$DFTB NDFTB = 2 DISP = UFF PARAM = MATSCI02 $END
ob3 ($\omega = 0.3$)	protein	LC-DFTB2/D3(BJ)	$DFTB NDFTB = 2 LCDFTB = .T. PARAM = OB3W0PT3 EMU = 0.3 $END $DFT DC = .T. IDCVER = 4 $END

[a]The recommended dispersion model is specified after the slash, although only UFF may be used with PBC.

[b]In addition, $BASIS GBASIS = DFTB $END should be added. If there are detached covalent bonds, $FMOHYB should define DFTB hybrid orbitals.

The most convenient way to specify parameters is with the PARAM keyword in $DFTB, in which case the root part of the path where are parameters are stored is taken from the DFTBPAR variable in gms-files.csh, by default, auxdata/DFTB. Alternatively, one can specify the path for each pair of atoms in $DFTBSK.

When needed, basis sets can be read in DFTB either by providing an empty $DFTBAO $END group, so that AO basis sets are taken from the same directory as parameters, or by specifying file names explicitly in $DFTBAO. Basis sets in DFTB are of the Slater type with spherical angular components. Providing basis sets in a DFTB calculation does not change the results; it only permits calculation of additional properties (electron density on a grid).

To read a new parameter set in PARAM, the directory name should be in capital letters, at most 8 characters long. The file name for AOs should begin with a lower case "wfc-" appended to the parameter name (upper case) and .hsd extension. For a DFTB parameter set, there is a readme file in the corresponding directory, where the details of input options and required citations are listed.

There is a damping (DAMPXH = .T.) developed for treating electrostatic interactions with hydrogen atoms, which is normally used in DFTB3, and may be turned on optionally in DFTB2. Using this damping can result in unphysical charges on metal ions (a negative charge for Na^+). It is advised to turn off damping for metal ions, if any, with ISPDMP (*i*). The index *i* should be chosen based on $DATA. If Na is the second element in $DATA, then ISPDMP(2) = 0 disables damping for all Na atoms (DFTB pedants might say in all Na species).

Several features of DFTB are unique among QM methods: it is possible to use PBC (Γ-point only), there is an exact analytic gradient for AFO, and exact analytic Hessian in vacuum. On the other hand, many things are not possible with DFTB: EFP solvent, multilayer FMO, analytic PCM Hessians, open-shell and excited state calculations.

In making DFTB input files, there is one point of concern: the atom names in $DATA should be element names without additional indices (Section 2.6.1), for example, one should use "H" rather than "H-1", etc. Some useful options for DFTB are listed in Table 50.

Table 50. Useful options in $DFTB.

Value	Meaning
DISP = UFF	Use UFF dispersion.[a]
MODGAM = 13	Accelerate DFTB (requires a large memory for storing γ (Γ) matrices).[b]
PBCBOX (1) = ...	Lattice vectors in PBC (specify 3 elements of the first vector **a**, then the second vector **b**, and the third vector **c**).[c]
LATOPT = 0	In PBC, fix the lattice vectors.
LATOPT = 1	In PBC, unconstrained lattice optimization.
LATOPT = 2	In PBC, constrain the cell to be orthogonal during its optimization.

[a]Other types of dispersion are chosen with $DFT DC = .T. IDCVER = *K* $END (for D3(BJ), K = 4).
[b]This option cannot be used for FMO0 and Hessians.
[c]When this option is defined, PBC is turned on.

4.1.4 *Interface with Molecular Mechanics (MM)*

The interface for combining FMO with force fields uses a specially adapted version 3.6 of Tinker distributed with GAMESS. Other versions of Tinker may not be used. In this interface a single GAMESS/Tinker executable is built and run as any other GAMESS. This specific version of Tinker is parallelized in a way compatible with GDDI.

The type of QM/MM that can be used with FMO is a mechanical embedding (no polarization of QM by MM), with the definition of link atoms following surface integrated MO/MM (SIMOMM). Making FMO input files for use with Tinker and preparing parameters for Tinker are a difficult task. The FMO/MM interface can be used mainly for geometry optimizations, performed in an iterative two-step loop: one step of updating QM atomic coordinates followed by another step of a full optimization of MM atoms. Other jobs such as MD are not supported.

Tinker has a limit of the number of QM + MM atoms (MAXATM = 12000), and the number of link atoms (on the QM/MM boundary), MAXLNK = 100; they may be changed in GAMESS and Tinker source codes. Some Tinker optimizations tend to allocate more and more memory as they proceed, so that eventually memory may become a problem (unless a less voracious algorithm is chosen).

Covalent boundaries between QM and MM are allowed, so that a protein-ligand complex can be optimized, with the pocket treated by FMO and the rest with a force field. FMO/MM is an alternative to the frozen domain (Section 4.7), with the benefit that all atoms can be optimized together.

An example of an input section is shown in Table 51. Coordinates of all QM and MM atoms are in $TINXYZ in the Tinker format. QM

Table 51.   Force field related part of an FMO/MM input.

Input[a]	Comments
⌐ $TINXYZ	
328 FMO/MM example	Number of atoms
1 N3 4.595185 −1.827628 −2.371698 124 2 5 6 7	Coordinates of atom 1
2 CT 4.058126 −2.424688 −1.133000 134 1 3 8 9	Coordinates of atom 2
...	
328 O2 18.021721 14.916638 22.297832 112 306	Coordinates of atom 328
⌐ $END	
⌐ $LINK	
SIMOMM = .T.	Use SIMOMM
IQMATM(1) = 1,2,3,4,5,6,7,8,9,10,11,12,13,14,17,	QM region
18,19, 15,16,20,21,24,25,26,27,28,29,30,31,32,	(serial numbers of atoms)
33,34,35	
⌐ $END	

(Continued)

Table 51. (*Continued*)

Input[a]	Comments
␣$TINKEY	Each option on a separate line.
parameters /home/user/gamess/tinker/params/oplsaa	FF parameter file.
verbose	Increase output.
steepest-descent	Optimization algorithm.
␣$END	
␣$TOPTMZ	
FRZMM = .F.	Do not freeze MM atoms.
␣$END	

[a]Truncated, with some spaces removed.

atoms are chosen in IQMATM of $LINK. The file name for FF parameters, as well as other Tinker options, are specified in $TINKEY. Optimization options are in $TOPTMZ.

The FMO section of the input should be made assuming that atoms are numbered consequently from 1 (irrespective of their serial numbers in $TINXYZ). The FMO part is straightforward. The most difficult part is to prepare force field parameters (for a ligand).

Another possibility to do FMO/MM is to use force fields via the EFP interface (Section 3.1.2). EFP can be used to embody common force fields such as TIPnP. A rudimentary way of doing electronic embedding (QM/MM) is possible via fixed fragment charges (Section 5.2).

4.1.5 *Pseudopotentials*

A pseudopotential (a core potential) replaces some electrons. Core potentials are often used for heavy atoms such as Pt or Au, but one can also find potentials for atoms as light as carbon or oxygen. GAMESS has two types of core potentials, effective core potentials (ECP) and model core potentials (MCP). The latter have an extra term, enforcing the proper shape of valence orbitals in the core region.

A potential (PP in $CONTRL) and a basis set for it should be defined. Some of these for MCP are distributed in GAMESS in external files (which are read on the fly). It is possible to use core potentials for some atoms and all-electron basis sets for other atoms. For multilayer runs, the same core potential for an atom is used in all layers; but basis sets may be layer dependent.

For detaching covalent bonds at a BDA using a core potential, hybrid orbitals for that atom should be generated with the core potential and its basis set. For example, for a C–O bond with a core potential for C (BDA), hybrid orbitals have to be generated for C (see Section 4.11).

Next, specific points for making input files are given. The molecular system is zinc cation in explicit water. It should be noted that if a metal that forms dative bonds is made a separate fragment, then PRTDST(3) = 0.4 in $FMOPRP should be added to the input. This is because interfragment (Zn–O in case of solvated Zn) bonds are so short that the program detects them as covalent bonds between fragments with no corresponding entries in $FMOBND. To turn off the check, the PRTDST(3) option is set.

It may be necessary to remove layer specifications in atom names in $DATA to enable matching of names in $DATA and in a core potential in some cases, i.e., instead of "C-1" use "C" in $DATA.

4.1.5.1 *Effective Core Potentials (ECP)*

A general example of a user-specified ECP is shown in Table 52, with an ECP only for Zn. The order of atoms in $ECP should be the same as in $DATA, and each element in $DATA should appear in $ECP. There may be 100 water molecules, but in $ECP only one O and one H are put. For multilayer runs, each element in $ECP appears once per layer as in $DATA, in the same order. However, the ECP for an atom in all layers should be the same.

The actual potentials and basis sets can be conveniently obtained from the database,[11] that can generate them in the right format for GAMESS, so the user does not have to learn format details.

Table 52.　ECP input options for Zn in explicit water.

Input	Comments
␣$CONTRL PP = READ $END	Manually defined ECP.
␣$DATA	
ECP example for solvated Zn	
C1	
O 8	
N31 6	All-electron 6-31G for O.
H 1	
N31 6	All-electron 6-31G for H.
Zn 30	ECP basis set for Zn
S 2	(truncated).
1 0.7997　−0.6486112	
2 0.1752　1.3138291	
...	
␣$END	
␣$ECP	
O　NONE	No ECP to be used for all oxygens.
H　NONE	No ECP to be used for all hydrogens.
ZN　GEN　18　3	Manually defined potential for Zn
5　----- f-ul potential -----	(truncated).
−18.0000000　　1　386.7379660	
−124.3527403　2　72.8587359	
−30.6601822　　2　15.9066170	
−10.6358989　　2　4.3502340	
−0.7683623　　2　1.2842199	
...	
␣$END	

There are also some built-in ECPs with predefined basis sets. For example, PP = SBKJC in $CONTRL and GBASIS = SBKJC in $BASIS can be used.

4.1.5.2 *Model Core Potentials (MCP)*

Whereas for ECPs built into GAMESS there are only small basis sets, built-in MCPs come with good quality basis sets. MCP is defined in $MCP. An example of an input is shown in Table 53.

Table 53. MCP input options with manual definitions for Zn in explicit water.

Input	Comments
⌴$CONTRL PP = MCP $END	Core potential is MCP.
⌴$DATA	
Manual MCP example for solvated Zn	
C1	
O 8	
N21 3	All-electron 3-21G for O.
H 1	
N21 3	All-electron 3-21G for H.
Zn 30	
MCP READ	Manually defined MCP.
S 4	MCP basis set for Zn
1 366.2133500 0.16998540	(truncated).
2 13.8712140 −0.64669630	
3 2.7859173 0.56513980	
4 1.0813192 0.64977480	
...	
⌴$END	
⌴$MCP	
MCP for Zn	
Zn	Element name.
6 2 10 1 7	MCP for Zn.
20.00(5F14.7)	(truncated).
0.2353145 0.0789866	
0.0004424	
...	
⌴$END	

The input format for MCP is slightly shorter than for ECP. There is no need to specify explicitly that O and H do not use an MCP: one defines that Zn has an MCP and that is it. There is no need to match the order of atoms in $DATA for MCP.

Table 54.   MCP input options for Zn in explicit water.

Input	Comments
␣$CONTRL PP = MCP $END	MCP.[a]
␣$BASIS GBASIS = MCP-TZP $END	Triple-ζ basis set.
␣$DATA	When GBASIS is defined, no basis set
Built-in MCP example for solvated Zn	may be given in $DATA.
C1	The use of GBASIS for an MCP also
O 8	means that built-in potentials are
H 1	used.
Zn 30	
␣$END	

[a]Both a core potential PP = MCP and a basis set for it (GBASIS) should be defined.

An input is much simpler for the built-in basis sets and potentials (Table 54). The problem is that built-in MCP basis sets for Zn are large, and to use them, one should retort to one of the options for such basis sets (Section 5.3).

4.2 Unrestricted (UHF and UDFT) and Open-shell (ROHF) Methods

ROHF and UHF can be used to describe radicals. The former preserves the electron spin, whereas the latter has an inherent spin contamination (meaning that instead of a pure singlet $S = 0$ or a triplet $S = 1$, some fractional S "in between" may be obtained). UHF can describe a single bond breaking. UDFT is useful for describing transition metals with partially filled d-orbitals, in lieu of the theoretically more sound but expensive MCSCF.

UHF and UDFT have a similar functionality in FMO. ROMP2, ROCC, and UMP2 are available, with an approximate gradient for ROMP2. For ROHF or UHF (UDFT) the SCF type for each fragment can be specified, possibly with multiple open-shell fragments. It is not allowed to mix ROHF and UHF fragments, and it is impossible to mix non-DFT and DFT fragments in the same layer. For a Hessian calculation or in UTDDFT, only one open-shell fragment is permitted.

As an example, a typical scenario is explained for ROHF. SCFTYP(1) = ROHF in $FMO is used to declare that layer 1 is of

ROHF type. SCFFRG defines the SCF types for each fragment, as in SCFFRG(1) = RHF,ROHF,RHF,RHF (fragment #2 is ROHF, the rest RHF). SCFTYP in $CONTRL should also be set to ROHF.

The spin S should be set for each fragment in MULT of $FMO, as a $(2S + 1)$ value (1 for a singlet, 3 for a triplet, etc.). The MULT value in $CONTRL is ignored.

In FMOn, n-mers are computed with the open-shell method if any of n fragments is of an open-shell type, and with RHF (or RDFT) otherwise. When two or more fragments in an n-mer are open-shell, the highest spin coupling is chosen (combining a triplet with a triplet results in a quintet, according to the angular momentum addition rule). RHF fragments should be singlets, whereas ROHF and UHF fragments may have any multiplicity.

Open-shell calculations, especially UHF, are often difficult to converge, and the user may have to retort to some of the advanced SCF options (Section 4.10). A UHF fragment may take much longer than an RHF fragment, and an appropriate setup may be needed for parallel load balancing (Section 5.10).

A peculiar point of UHF singlets is that the default initial orbitals are of a pure singlet character (α MOs are equal to β MOs) and UHF starting from them typically stays a pure singlet (same as RHF). To break the spin symmetry and get a UHF solution, MIX = .T. in $GUESS can be added.

4.3 Electron Correlation

In the wave function approach to treating electron correlation, the energy is commonly expressed in terms of molecular orbitals. A transformation of integrals from the atomic to molecular orbital basis is needed. If this step is done for all orbitals at the same time, it may require more memory than is available. To learn about memory usage for correlated methods, read the output file.

A transformation can be divided into multiple passes. Think of it as having to carry a bucket of water from the well to your kitchen several times if the bucket is too small. In this case it may be possible to accelerate the calculation by providing more memory (invest in a bigger bucket).

4.3.1 *Core Inconsistency Problem*

In MP2 or CC, it is common to exclude some low-lying core MOs from correlated calculations, by assigning them to frozen (uncorrelated) core. While normally this is reasonable and works well, there is an FMO-specific problem related to this.

A core inconsistency problem is here illustrated for a cation solvated in explicit water. This problem is the core "shadow" of the charge instability problem of valence electrons (Section 4.10.1). Let us take a specific example of K^+, for which the frozen core is 1s2s2p3s3p (9 orbitals are mechanically assigned as the core for K^+, counting from the bottom of MO energies). However, in a K^+-water dimer, $\sigma(O-H)$ orbitals of water have a lower energy than 3p(K). Therefore, in the monomer K^+, 3p is correlated, and in the dimer $K(H_2O)^+$, it is not (because $\sigma(O-H)$ takes its place). Hence, the MBE of the correlated energies results in a misaddition of contributions.

A solution is to promote 3s and 3p of K from frozen to correlated core. To do this, the option NUMCOR(19) = 5 in $CONTRL can be used, which sets 5 orbitals (1s2s2p) of K ($Z = 19$) as uncorrelated core, rather than the default of 9. There is no automatic check for this problem, and it can only be detected by a very attentive user, who questions unexpectedly large dispersion contributions.

Another problem can occur with core potentials. For example, SBKJC ECP for Zn^{2+} leaves no valence electrons (an electron-less fragment), and the electron correlation for such a fragment is not computed.

4.3.2 *Møller-Plesset Perturbation Theory (MP2)*

MP2 is a good method for describing hydrophobic (dispersion) interactions, and it is widely used in biochemical applications. FMO-MP2 is affordable and for typical fragments containing 1–2 dozens of atoms it is not much more expensive than FMO-HF.

In GAMESS, only the second order can be used (MP2), set with MPLEVL = 2 either in $CONTRL or in $FMO. There are several restricted MP2 algorithms available, selected by CODE in $MP2.

Restricted closed and open-shell as well as unrestricted MP2 calculations are possible.

For CODE = IMS, a substantial I/O is done, but the code is very fast and has a reasonable memory consumption. I/O can be accelerated by using SSD or an electronic disk (Section 5.8.1).

For CODE = DDI, no I/O is done and all data are stored in shared memory invoking a huge amount of short parallel communications. This can be very slow on networks with a large latency such as Gigabit and its younger siblings (to reduce the parallel overhead, one node per group can be used in GDDI). The problem of CODE = DDI is the need to provide enough shared memory in MEMDDI.

IMS and DDI codes yield the same numerical results. The other choice, CODE = RIMP2, is an approximation to MP2, using the resolution of the identity (RI), with an energy slightly different from regular MP2. RI-MP2 can offer modest savings over regular MP2 for small fragments, and a substantial acceleration for large fragments.

RI-MP2 code normally uses both shared and replicated memory. The problem of this code is that it requires an auxiliary basis set, tailored to the main basis set. In the absence of an exact match, a basis set of a similar quality is used. For example, there is no predefined auxiliary basis set for 6-31G*, so as an ersatz, the auxiliary basis set made for cc-pVDZ can be used. It is possible to provide an external file containing auxiliary basis sets, similar to a file for main basis sets (these two external files are different). The option to do that is EXTCAB = .T. in $AUXBAS, and the file location is defined as EXTCAB in gms-files.csh.

In a gradient calculation an MP2 density matrix is calculated, which can be used to compute properties, such as dipole moments. To get MP2 density without the gradient, MP2PRP = .T. in $MP2 can be set.

Most MP2 codes print both the regular and spin component-scaled MP2 (SCS-MP2) energies. GAMESS uses analytically derived SCS-MP2 parameters, and there is no way to alter them.

Although two MP2 energies (regular and SCS) are computed, only one is picked to be used in geometry optimizations, etc. By default, the regular MP2 energy is picked; by selecting SCSPT = SCS in $MP2, the SCS-MP2 value is used instead. The same applies to pair interactions in FMO: to compute them with SCS-MP2, SCSPT = SCS can be set.

4.3.3 *Coupled Cluster (CCSD(T))*

Coupled cluster is a high-level method, with a variety of orders of incorporating the electron correlation, and a systematic way of improving the accuracy. CCSD(T) may be able to deliver high chemical accuracy for closed-shell molecules. For systems with a single broken bond other types of CC, such as CR-CC(2,3) (chosen with CCTYP = CR-CCL), are suggested as a better alternative. To select CC, CCTYP in $CONTRL or in $FMO can be set.

Various types of CC energies can be computed in GAMESS but only CCSD and CCSD(T) are parallelized in the regular version (RI-CC is available with OpenMP). CC codes use a lot of I/O and memory.

The parallel CCSD(T) code uses all three types of memory (Section 2.3.3). EXETYP = CHECK can hopefully give hints to the required memory, but in FMO one would have to collect these numbers from all GDDI groups and find the maximum of them among all fragments, dimers, and trimers.

Both restricted closed-shell CC and open-shell ROCC are available. There is no analytic gradient for any CC. CC methods have a hierarchy; in a CCSD(T) calculation, MP2 and CCSD energies are also computed. PIEs are calculated with the highest level, but some summary of total properties is provided for lower levels. The optional CC methods (provided by MSU) cannot be used with FMO.

4.4 Multiconfigurational Self-consistent Field (MCSCF)

MCSCF can be used to describe systems with partially occupied orbitals, for example, d-orbitals in transition metals. Only one fragment can be computed with MCSCF, and for all other fragments RHF is used. Dimers containing the MCSCF fragment are computed with

MCSCF, whereas other dimers (composed of two RHF fragments) are computed with RHF.

In an MCSCF calculation, it is necessary to define the number of core (doubly occupied), active (partially occupied), and virtual (unoccupied) orbitals, based on chemical considerations. It is often difficult, because ideally one would put all quasidegenerate orbitals in the active space (such as all π,π^* orbitals), and the cost of calculations grows exponentially with the active space size.

There are two varieties of MCSCF, state specific (SS) and state averaged (SA). In the former, a single CI root (state) is chosen. In the latter, several states are computed and their density is averaged with user-specified weights. SA-MCSCF can be used when several states are of interest. The method is often employed for geometry optimizations and computing properties involving multiple states. The gradient of SA-MCSCF is difficult, requiring CP equations.

For any MCSCF calculation, a CI basis has to be chosen. There are two choices: state configuration functions (CSF) or Slater determinants. CI vectors are expanded in terms of one of these bases. The choice is made via CISTEP in $MCSCF, set to GUGA for CSFs or ALDET for determinants. Each CSF has a definite spin, but determinants in general have no definite spin (CSFs are linear combinations of determinants). A CI Hamiltonian in the basis of CSFs has a smaller dimension than for determinants, and all eigenvalues for a CSF Hamiltonian have the same spin. For determinants, CI states can have various spin multiplicities, which may be useful for averaging states in MCSCF. The disadvantage of the CSF code is that it uses I/O extensively, so that large active spaces are easier to accommodate with determinants.

To set up an MCSCF calculation (here, the determinant CI is described, with the group of $DET), one has to define the number of core orbitals as NCORE and the number of active orbitals as NACT. The number of virtual orbitals is usually determined automatically (all other orbitals). The number of electrons in the active space is set as NELS.

The number of computed states is defined as NSTATE, and their weights are given in WSTATE. If PURES is .F., then states of different multiplicity may be averaged. In FMO, NCORE should be set for the MCSCF monomer, and it is automatically adjusted for dimers. The active space (NACT and NELS) is the same in all monomers and dimers.

There is a way to accelerate MCSCF by using a two-layer run, with RHF and MCSCF in layer 1 and 2, respectively. This limits MCSCF dimers to pairs in layer 2. In the most economic case, one can avoid all MCSCF dimers by placing only the MCSCF fragment in layer 2.

Doing MCSCF calculations requires a prepared set of initial orbitals, provided by the user. There are two ways of reading initial MOs, using a binary restart file (F40) or text MOs in the input file.

Using a restart via F40 is the older way, which is rarely used. Typically, first FMO-RHF calculations are done with MODORB = 3, and F40 is saved. Then an orbital conversion run is performed via a binary restart (Section 5.5.4), whose purpose is to transform F40 from the RHF format to the MCSCF format, and to reorder orbitals at the same time. This is done with MAXAOC = M in $FMOPRP ($M$ is the maximum number of AOs per fragment). MOs can be reordered with $GUESS GUESS = MODAF NORDER = 1 IORDER(20) = 21,20 $END (to swap orbitals 20 and 21). The result of an orbital conversion is stored as a F30 file. This file can be renamed to F40 and used in an MCSCF calculation with a binary restart.

The alternative way is to use the general method of reading initial orbitals, not limited to MCSCF (see Section 5.5.3). First, an FMO-RHF calculation is done with NPUNCH = 2 in $SCF to have MOs written to the DAT file (for reducing the output, NPUNCH = 0 can be combined with NOPFRG(n) = 1 in $FMO, where n is the ordinal number of the MCSCF fragment). Then the MOs in $VEC from RHF written to the DAT file can be copied to a $VEC1 group in an MCSCF input, and manually reordered if needed. To manually reorder the orbitals, the user has to know the format of $VEC. For MCSCF, it is important that all MOs be provided, all occupied and all virtual (for RHF, occupied orbitals suffice).

MOs $C_{\mu i}$ are written to $VEC starting from MO $i = 1$ spanning all AOs μ, followed by MO $i = 2$, etc. The initial two digits in each entry of $VEC are mod($i$,100), that is, the last two digits of i. So "99" is followed by "00", and then by "01". To reorder two MOs, one has to swap sets of data for desired MOs and update the initial two digits.

The convergence of MCSCF is often difficult, and various options are available (Sections 4.10.5 and 4.10.6). There is an approximate gradient (Section 2.6.2.8), but no Hessian. Many features such as EFP or PIEDA cannot be combined with MCSCF, and there is no FMO3-MCSCF.

4.5 Electronic Excited States

4.5.1 *SCF Approaches (ROHF, UHF, and MCSCF)*

Excited states can be computed self-consistently in SCF. For example, if the ground state is a singlet, the lowest triplet may be an excited state, which can be calculated in SCF (ROHF or UHF). MCSCF is an even more convenient framework. In some systems a triplet is the ground state.

Because space symmetry is not used in FMO, one cannot select say A_2 or B_1 states in C_{2v}. The only way to distinguish states explicitly is by their spin. Using SCF to get excited states is normally limited to the lowest state in each multiplicity (SA-MCSCF offers more flexibility, see Section 4.4). MP2 may be used to add electron correlation.

Multiple chromophore fragments are possible with ROHF, UHF, and UDFT, although not all tasks may use them (for example, Hessians cannot). See Section 4.2 for more details.

4.5.2 *Single CIS/TDDFT Chromophore Fragment*

Excited states can be calculated using CI with single excitations (CIS) or time-dependent DFT (TDDFT). For the latter, there is also its unrestricted variation. It is possible to do time-dependent HF (TDHF) calculations, which are technically very similar to TDDFT. They are accomplished by specifying DFTTYP = NONE in $CONTRL. $TDDFT applies to any TD method, including TDHF. CIS is an approximation to TDHF.

For CIS, CITYP = CIS in $CONTRL is used, and some details of CIS calculations are specified in $CIS such as the multiplicity MULT of the excited state of interest. Higher orders (FMO2 and FMO3) may not be used for CIS, and there is no analytic gradient.

FMO-TDDFT can be chosen with TDDFT = EXCITE for any order in FMO, and there is an analytic gradient for singlet states in vacuum. Similarly to MP2PRP, there is a TDPRP option to get density of an excited state (chosen with IROOT in $TDDFT) for the purpose of calculating some properties (note that in any gradient calculation, an excited state density is always computed). There are two different ways of running FMO-TDDFT: the first is suitable for one chromophore fragment and the second is used for multiple chromophores that are coupled in a model Hamiltonian.

First, the case with one chromophore is considered. This can be a photoactive molecule in an explicit solvent, or a chromophore fragment in a protein. For this approach, one fragment is selected as the chromophore, and for all other fragments the ground state is calculated. For the monomer, dimers, and trimers containing the chromophore, excited state calculations are done, and the rest are calculated in the ground state.

However, there is a difficulty. It is necessary to match the excited state in the chromophore fragment to an excited state in a dimer or trimer. Suppose one is interested in an excited singlet state #1 in fragment 3. It should be determined which excited state in dimer 3,1 corresponds to state #1 in fragment 3. It is conceivable that it may not be state #1 but another state. This search for a matching state is done automatically. The principle for matching states is to consider the amplitudes and the overlap of molecular orbitals in the monomer and dimer. In most cases, this matching works well, especially for low-lying excitations. For high excited states, the matching may not work well (or at all) for the physical reasons of the nature of the excited state being too different (in the monomer and dimer). For each dimer, a confidence measure of matching is printed (Table 55), which should ideally be 1 (that is, 100%). The threshold EXFID in $FMO can be used to neglect contributions with a confidence less

Table 55. Sample of a FMO2-TDDFT calculation results.

STATE 1, EXCITATION ENERGY =   4.260 eV (2 corrections out of   2)

IFG	NAME	RIJ	confidence,%	contribution,eV [a]
1	HCOH	0.000	100.0	4.180
2	WAT001	0.786	91.2	0.021
3	WAT002	0.999	91.2	0.059

STATE 2, EXCITATION ENERGY =   6.488 eV (2 corrections out of   2)

IFG	NAME	RIJ	confidence,%	contribution,eV [a]
1	HCOH	0.000	100.0	9.585
2	WAT001	0.786	2.3	−1.700
3	WAT002	0.999	1.2	−1.397

[a]Contribution means either ω_L^i (IFG is L = 1) or $\Delta\omega_{IL}^i$ (IFG is $I \neq L$), see Eq. (8). The excited state i here is either 1 or 2.

than EXFID. A sensible choice is 0.5 (50%), whereas the default is 0 to add all contributions.

In Table 55, the chromophore is fragment L = 1 (see Eq. (8)), whereas fragments I = 2 and 3 are solvent molecules. The confidence for the chromophore itself is always 100%, because it is the monomer contribution ω_L^i. For pairwise contributions (solvatochromic shifts due to individual solvent molecules) $\Delta\omega_{IL}^i$, the confidence for the first excited state is high (91.2%). For the second excited state, the confidence for pairwise contributions is very low, 1–2% (the default EXFID was used, so that these suspicious contributions are added). For the second state, only the monomer energy   = 9.585 eV is meaningful.

To help the program find the right state, the user can consider specifying more states to be computed than needed; for example, if the state of interest is #2, one can ask for 5 states, because in one of the dimers it may be state #3 or #4 that matches #2 in the chromophore. It should be noted that the computational time substantially increases with the number of states.

An excitation energy $\omega^{FMO2,i}$ is the sum of contributions of fragment L, and N–1 dimers IL (of N–1 dimers, most may be computed quickly with the RCORSD approximation, and only close pairs are done with TDDFT). On the other hand, to get $E^{FMO2,i}$, all ground state

Table 56.   Options for FMO*n*-TDDFT calculations.[a]

Option	Meaning
IEXCIT(1)	ID of the chromophore fragment.
IEXCIT(2)	*n*.
IEXCIT(3)	0 for excitation energy $\omega^{FM02,i}$ (fast), 1 for excited state energy $E^{FM02,i}$.
IEXCIT(4)	Monomer-dimer matching method, usually 2.

[a]For FRET, the usage is described in Section 4.5.3.

DFT dimers should also be computed. Obtaining an energy gradient requires a calculation of $E^{FM02,i}$, whereas the cost of a single point energy calculation can be largely reduced by calculating only $\omega^{FM02,i}$ (omitting all DFT dimers).

The input options are summarized in Table 56. It is possible to do an FMO1-TDDFT calculation combined with FMO2-DFT, neglecting pair contributions to excitation energies, by setting IEXCIT(2) = 1 in $FMO. An example of options for a gradient calculation is IEXCIT(1) = 4,2,1,2, where the chromophore fragment is 4, 2 selects FMO2, 1 is set to compute the full excited state energy, and 2 is the dimer matching method.

The state-matching algorithm does not work well for a cluster of identical chromophores. If one is to pick only one molecule as the TDDFT fragment, then in a dimer calculation there will be two quasidegenerate excited states coupled. Each of the two states matches the state in the monomer with a ≈50% confidence, and the MBE is hard to apply. In this case, FRET should be used.

4.5.3 *Multiple TDDFT Chromophore Fragments (FRET)*

For multiple chromophores, the alternative way of computing excited states is based on the Förster resonance electron transfer (FRET) theory, combined with FMO1 (or FMO0). In this approach, a set of chromophore fragments is selected by the user. For each of them, an excited state is calculated. Then the coupling of the excited states is estimated with a special model, based on the electrostatic interaction between transition densities plus a screening contribution if PCM is used. This produces a symmetric Hamiltonian matrix $\mathbf{H}^{\text{FRET}}$, whose linear size K is equal to the number of chromophore fragments.

$$\mathbf{H}^{\text{FRET}} = \begin{bmatrix} \omega_1 & H_{12} & \dots & H_{1K} \\ H_{21} & \omega_2 & \dots & H_{2K} \\ \dots & \dots & \dots & \dots \\ H_{K1} & H_{K2} & \dots & \omega_K \end{bmatrix} \tag{75}$$

where ω_I is the excitation energy of an exciton in chromophore I, and H_{IJ} is the excitonic coupling between chromophores I and J. H_{IJ} can be computed with transition densities or in the dipole approximation; in PCM H_{IJ} includes solvent screening. $\mathbf{H}^{\text{FRET}}$ is diagonalized, resulting in delocalized excited states.

The input option for FRET is IEXCIT(1) = −1,1,2,0 in \$FMO (these 4 options are fixed for any FRET) and IACTFG in \$FMO defines the excited state for each fragment. For example, IACTFG(1) = 0,1,1,0,2 means: do not include fragments 1 and 4 in FRET; take the lowest (first, 1) excited states of fragments 2 and 3, and the second (2) excited state of fragment 5. The FRET matrix will have the size of 3 × 3.

4.5.4 *TDDFT Calculations in Solution (PCM)*

TDDFT calculations can be combined with PCM for the energy (but there is no analytic gradient). First, ground state DFT/PCM calculations are done, followed by TDDFT/PCM to get excited states, in which induced solvent charges are computed for the transition density.

There are two approaches to TDDFT/PCM, equilibrium and non-equilibrium. The first assumes that the excitation is relaxed in equilibrium, which is appropriate for optimizing excited state structures and emission spectra (not implemented for FMO). In this case, the bulk dielectric constant ε of the solvent is used to obtain PCM charges due to TDDFT transition density.

The second, non-equilibrium approach is for vertical excitations and adsorption spectra. In this case an optical dielectric constant ε_{opt}

is used in TDDFT/PCM, specified as EPSINF in $PCM and NONEQ = .T. is set in $TDDFT. PCM may also be used to describe a protein medium, in which case suitable ε and ε_{opt} for proteins can be used.

4.6 Multiple Layers

The multilayer FMO (MFMO) scheme is illustrated in Figure 16 for the case of two layers. Each fragment is assigned to either lower layer 1 or higher layer 2. For instance, layer 2 may be the active site in a protein, and layer 1 is the rest of the protein. For each layer, a basis set and a wave function are chosen (Section 2.6.1). DFTB may not be used in a multilayer calculation, otherwise all supported QM methods may be employed.

First, all fragments from both layers are calculated at the level of layer 1, as shown by the uniformly yellow circle in the middle of Figure 16, which explicitly includes the higher region. In other words, the first step is to do a regular FMO calculation at the level of layer 1 for all fragments. The only difference, indicated by the prime in 1′, is that dimers *IJ* are not computed if both *I* and *J* are in layer 2.

Secondly, fragments in layer 2 are recomputed using the QM method and basis set of layer 2. These fragments are immersed in the combined embedding due to fragments in layer 1 (fixed in this step, light yellow) and layer 2 (self-consistently updated, blue). This step is like a regular FMO calculation for fragments in layer 2, but with an extra fixed embedding potential from layer 1.

The energy in multilayer FMO is written like the general expression in Eq. (3), but for each fragment in this expression, the energy of

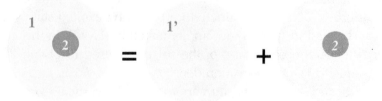

Figure 16. Overall scheme of multilayer FMO, for two layers 1 and 2.

the highest computed layer is used. For example, if *I* and *J* are in layer 2, then the energy of dimer *IJ* is for layer 2; if either fragment is in layer 1, then the energy of *IJ* is for layer 1. Likewise, more layers can be treated (the maximum number is 5).

4.7 Frozen Domain (FD)

The frozen domain (FD) is an economic way to optimize geometry of a selection of atoms and to calculate the Hessian. A typical example is an active site in a protein binding a ligand, or an enzyme catalyzing a chemical reaction.

The idea of this approach is to divide all fragments into frozen and active domains. The atoms in the frozen (**F**) domain stay fixed, their electronic state is computed for the initial geometry, and it does not change for other geometries. The atoms in the active domain (**A**) are allowed to move, and their electronic state changes. There is another domain, called polarizable buffer (**b**), which surrounds the active domain. The atoms in **b** are fixed, but their electronic state is allowed to change at each step in a geometry optimization. The union of **A** and **b** is the polarizable domain **B** (Figure 17).

A FD calculation is accomplished by setting up **F** and **B** as layers 1 and 2, respectively. In each layer (see Section 4.6), one can select a wave function and a basis set. A summary of domains is shown in Table 57.

Although technically a two-layer calculation, in the variational sense, FD is a single (second) layer calculation. The first layer is used as a fixed embedding for the second layer. The electrostatic

Figure 17. Two-layer FMO used to define frozen domain (**F**), active domain (**A**) and polarizable buffer **b**. **F** and **B** = **A** ∪ **b** form layers 1 and 2, respectively.

Table 57.   Frozen domain summary.

Domain	Layer	Atoms	Electronic state	MXATM[a]
Frozen (**F**)	1	frozen	frozen	no
Buffer (**b**)	2	frozen	polarizable	yes
Active (**A**)	2	active[b]	polarizable	yes

[a]Whether the number of atoms is limited by MXATM or not.
[b]Some atoms may be frozen.

interactions between fragments in **F** and **B** explicitly contribute to the energy. Analytic gradients and Hessians can be computed for FD, and PCM may be used. RUNTYP for FD is limited to OPTIMIZE, SADPOINT, IRC, HESSIAN, and RAMAN (note that OPTFMO and ENERGY are not possible for FD).

There is an approximation, FD with omitted dimers (FDD). Without the FDD approximation, all dimers in **B** are computed. In FDD, only such dimers in **B** are calculated, for which at least one fragment is in **A**.

The size of **b** is chosen by the user. If there are covalent bonds between **A** and the rest of the system, **b** should include at least fragments covalently bound to **A**. **A** is chosen as the union of fragments containing the atoms to optimize. Typically, all fragments within a certain threshold R^b from **A** are selected as **b**, and the rest is assigned to **F**. The recommended threshold R^b is 3.9 and 6.5 Å for a neutral and charged **A**, respectively (a charged domain polarizes more strongly, so a larger threshold is used). The distance between **A** and **F** is calculated at each step in a geometry optimization, so the user may decide upon a redefinition of domains if the buffer **b** between **A** and **F** becomes too thin (Figure 17).

There are several special cases of defining domains. One is to have no **F**, which makes sense only for FDD, where many dimers are not computed in **B**. Another is to have a single fragment in **A**, for example, a ligand, which is especially fast with FDD.

To set up an FD calculation, an input file for two layers should be made (Section 4.6), which defines **F** and **B** domains. A domain is set up by listing fragments in **A** as IACTFG of $FMO. Full FD is chosen

Table 58. Sample header for an FDD calculation.

Input	Comments
␣$CONTRL RUNTYP = FMOHESS $END	A Hessian calculation.
␣$STATPT IACTAT(1) = 1,−7 $END	List of active atoms in **A**.
␣$FMO MODFD = 3	FDD is chosen.
IACTFG(1) = 1,8	Fragments 1 and 8 form the active domain **A**.
NLAYER = 2 LAYER(1) = 2,2,1,1,1,1,1,2 $END	Fragments 1,2,8 form the
␣$DATA	polarizable domain **B**.
FDD SAMPLE INPUT FOR (H2O)8	
C1	
H-1 1	
STO 3	Layer 1: STO-3G for hydrogen.
O-1 8	
STO 3	Layer 1: STO-3G for oxygen.
H-2 1	
N21 3	Layer 2: 3-21G for hydrogen.
O-2 8	
N21 3	Layer 2: 3-21G for oxygen.
␣$END	

with MODFD = 1, whereas FDD is selected with MODFD = 3 in $FMO.

The list of atoms in **A** that should be optimized is specified as IACTAT in $STATPT. It is possible to freeze some atoms in **A**, for example, one may like to optimize the positions of hydrogen atoms only. The format of IACTAT is similar to INDAT: ranges may be used, for example, "1 4 −6" defines atoms 1, 4, 5, and 6 as active. An example of a sample header for an FDD calculation is shown in Table 58.

4.8 Molecular Orbitals and Their Energies (LCMO)

Sometimes it is useful to draw molecular orbitals for some fragment, dimer, or trimer, in particular, for excited states and MCSCF. In order to plot MOs, it is necessary to set up the option NFRND = 2 in $FMO to write out the basis set for each fragment, and also use NPRINT = 7 in $CONTRL. NPREO in $SCF can be used to reduce the number of printed orbitals.

With these options, an output will have marked sections for each fragment (if GDDI is used, output files should be collected from all groups, see Section 2.7). The user can cut sections for fragments of interest and store them as separate text files, one per *n*-mer. They can be read into programs that can plot MOs using GAMESS output, for example, MacMolPlt. DFTB orbitals cannot be plotted in this way because AOs in DFTB are of the Slater type.

Computing molecular orbitals for the whole system is a more challenging task. There are two approaches for it, FMO/F (F stands for the Fock matrix) and linear combination of molecular orbitals (LCMO). For FMO/F, Fock and overlap matrices from the DAT file can be processed in an external program. On the other hand, LCMO builds and diagonalizes the total Fock matrix inside of GAMESS.

Both FMO/F and LCMO can be used with HOP and AFO, but only with closed-shell methods (RHF, RDFT, or DFTB). PCM may be used. For PBC, only the Γ-point can be computed, so it is impossible to obtain a zone structure with FMO (no density of states for PBC).

LCMO works as follows. The user has to select some molecular orbitals of fragments, which are coupled in the total Fock matrix, the diagonalization of which yields MOs and their energies for the whole system. A scheme of LCMO is described in Table 59.

Table 59. Scheme of LCMO calculations.

Step
1. Select a set of fragment MOs (for example, HOMO and LUMO of each fragment), $\{\varphi_i\}$.
2. Build the total Fock matrix F_{ij} in the basis of $\{\varphi_i\}$ using the MBE and a projection technique applied to Fock matrices of monomers, dimers, and, optionally, trimers.
3. Diagonalize F_{ij} and obtain its eigenvalues and eigenvectors expanded in the basis of $\{\varphi_i\}$.[a]
4. Transform MOs to the regular AO basis (μ), producing expansion coefficients $C_{\mu i}$.

[a]When there are detached bonds between fragments, there may be linear dependencies, which are removed via a transformation to a linearly independent basis (warnings are printed).

Table 60. Packed NLCMO(1) option in $FMOPRP for LCMO calculations of MOs and their energies.

Value	Meaning
1	Do an LCMO calculation and print orbital energies.
2	Use exchange in the embedding (LCMOX).
4	Print MOs.
8	Use normal approximations.
16	Print the LCMO Fock matrix F_{ij}.
32	Print the overlap matrix S_{ij}.

There are two variations of LCMO, regular LCMO and LCMOX. They differ in the embedding potential: in the former, the embedding potential is Coulomb; in the latter, it is Coulomb + exchange (as in regular HF). LCMOX is more accurate for FMO2. However, apart from some minor numerical artefacts of approximations, LCMOX and LCMO are identical for FMO3, because the embedding potential cancels out. LCMOX is recommended for FMO2, and LCMO for FMO3, but for DFTB, only LCMO may be used.

Likewise, there is FMO/F with exchange in the embedding, denoted by FMO/FX and turned on with MOFOCK = 3 in $FMOPRP (FMO/F is chosen with MOFOCK = 1).

For doing LCMO calculations, three options are defined as NLCMO in $FMOPRP. The first is described in Table 60. The second is the number of occupied orbitals in each monomer, counting from the highest occupied molecular orbital (HOMO) down. The third is the number of virtual orbitals in each monomer, starting from the lowest unoccupied molecular orbital (LUMO) up. 0 has a special meaning to include all respective orbitals. To add HOMO,HOMO-1 and LUMO of each fragment in LCMOX, NLCMO(1) = 15,2,1 can be used. NLCMO(1) = 13,0,0 to include all orbitals in LCMO.

4.9 Properties on a Grid

It is possible to calculate electron and spin densities, as well as molecular electrostatic potential (MEP) on a 3D grid, for the whole system or individual fragments. To obtain electron density on a grid for DFTB, it is necessary to define AOs (Section 4.1.3). The resultant data on a grid are written to DAT file in the cube format. Currently, it is not possible to create a cube file for a molecular orbital.

Cube files are supported in many common GUIs. MEP can be used to predict the docking site for a ligand bound to a protein, by computing their potentials and finding a matching orientation to maximize the attraction.

Using Eqs. (73) and (74), the electron and spin densities of fragment X are defined as

$$\rho_X(\mathbf{r}) = \sum_{\mu\nu} D_{\mu\nu}^X \chi_\mu(\mathbf{r}) \chi_\nu^*(\mathbf{r}) \tag{76}$$

$$\rho_X^{spin}(\mathbf{r}) = \sum_{\mu\nu} D_{\mu\nu}^{X,spin} \chi_\mu(\mathbf{r}) \chi_\nu^*(\mathbf{r}) \tag{77}$$

where $\chi_\mu(\mathbf{r})$ denotes AO μ. These densities can be summed in a MBE. For example, the electron density is defined in FMO2 as

$$\rho(\mathbf{r}) = \sum_{I=1}^{N} \rho_I(\mathbf{r}) + \sum_{I>J} \left[\rho_{IJ}(\mathbf{r}) - \rho_I(\mathbf{r}) - \rho_J(\mathbf{r}) \right] \tag{78}$$

In FMO3, a sum over three-body corrections $\Delta\rho_{IJK}(\mathbf{r})$ is added. Likewise, the spin density is computed.

The electron density is normalized to the total number of electrons,

$$\int_{-\infty}^{\infty} \rho(\mathbf{r}) d\mathbf{r} = N^{el} \tag{79}$$

A useful property is the molecular electrostatic potential (MEP),

$$\phi(\mathbf{r}) = \sum_A \frac{Z_A}{|\mathbf{r} - \mathbf{R}_A|} - \int_{-\infty}^{\infty} \frac{\rho(\mathbf{r}')}{|\mathbf{r} - \mathbf{r}'|} d\mathbf{r}' \tag{80}$$

where A numbers atoms with nuclear charge Z_A and coordinates $\mathbf{R}_A$. MEP can also be defined for transition density in TDDFT.

4.9.1 *Total Properties*

A grid is defined in $GRID by three vectors, which do not have to be orthogonal (although they often are), and SIZE (the spacing between grid points). There are two ways to define the box of grid points: manually (useful to compute properties for a part of the system, such as a binding pocket), or automatically (a rectangular box surrounding the molecular system with some extra margin, called the padding).

The manual way is accomplished by providing ORIGIN, XVEC, YVEC, and ZVEC. ORIGIN is a vector of three Cartesian coordinates defining one corner of a "cube" (a parallelepiped in general). From ORIGIN, three vectors XVEC, YVEC, and ZVEC define three sides of the parallelepiped. For instance, ORIGIN(1) = 0,0,0, XVEC(1) = 10,0,0, YVEC(1) = 0,20,0, and ZVEC(1) = 0,0,30 define a rectangular cuboid with the size of 10x20x30 (the units for these parameters in $GRID are specified as UNITS, ANGS or BOHR).

The automatic way is chosen by $GRID XVEC(1) = 1, YVEC(2) = 1, ZVEC(3) = 1 $END and $FMOPRP GRDPAD = R^p $END. Here, XVEC, YVEC, and ZVEC in the shorthand notation define a rectangular box. The origin of the box, and the size along each of the three directions (x, y, and z) are defined automatically from the molecular geometry by finding two most separated atoms, and adding an extra "breezing" space of R^p (R^p is unitless, similar to FMO distances; p stands for padding). $R^p \times W_A$ is the geometric padding for atom A with van der Waals radius W_A. R^p is usually set to 2.

To calculate MEP, add 32 to MODPRP in $FMOPRP; for spin density, add 128. Both MEP and spin density require that electron density be computed (4 in MODPRP). For example, using MDOPRP of 52 = 4 + 16 + 32 will automatically select the grid window (16), and compute both electron density (4) and MEP (32). MEP is a much more expensive calculation than electron density because MEP decays very slowly (as inverse distance).

In PCM, a solvent screening is added to MEP. To soften the point charge effect from the discretized solvent surface, a Gaussian smearing of the PCM potential is applied, the exponent of which can be set as PCMGEX in $ELPOT.

There are two ways of storing grid data: as a replicated or a distributed array. The former is the default, which may take a large amount of memory (MWORDS). The latter is chosen by adding 512 to MODPRP, in which case the grid array is distributed in parallel (stored in MEMDDI).

4.9.2 *Fragment Properties*

It is possible to compute properties on a grid for individual fragments. This is done by specifying an array of 3 integers NGRID in $FMOPRP, defining the number of grid points in the x, y, and z directions. For using this option, 4 should not be added to MODPRP (4 is used for total properties only).

The grid box is determined automatically for each monomer separately using GRDPAD in $FMOPRP for padding. Fragments i, for which data should be computed, are selected by adding 4 to their NOPFRG(i). For example, NOPFRG(8) = 4 and NOPFRG(24) = 4 will produce data on a grid for two fragments, 8 and 24.

There is an alternative way of computing grid properties for individual n-mers. It is to set up $GRID as for non-FMO calculations, that is, with a full definition of XVEC, YVEC, ZVEC, ORIGIN, and SIZE. To get density, $ELDEN IEDEN = 1 WHERE = GRID OUTPUT = PUNCH $END can be used, and to get MEP, $ELPOT IEPOT = 1 WHERE = GRID OUTPUT = PUNCH $END is set. This way, grid properties are computed after every fragment calculation, which can flood the DAT file (not an issue for FMO0 calculations). In FRET, MEP can be computed from the transition density of fragments, for which 1024 is added to NOPFRG(i) for some fragments i.

4.10 Struggling with Convergence

A converger is an algorithm that determines how MOs are updated during SCF. In most cases, any converger will do just fine, but in difficult cases a skillful choice of convergers can be critical. Open shells, especially in unrestricted SCF, near degeneracies (partially filled d-orbitals, etc.), zwitterionic systems, DFT, and MCSCF often present significant, and at times, insurmountable problems. There can be several low-lying electronic states, and a different solution (state) can be obtained in SCF depending on the converger, especially in metallic and zwitterionic systems.

One should distinguish three different metallicities in FMO: (1) a small HOMO-LUMO gap in a fragment, (2) a small gap in a dimer or trimer, and (3) a collective small gap in the total system. The former situation may be solved by a judiciously chosen SCF converger (or switching to a more appropriate QM method); this problem is not really specific to FMO. The second type is related to the charge instability in FMO, described next. The third type of metallicity, called collective, is related to the quantum confinement effect. It arises as a result of collective efforts of many fragments together, but not in individual n-mers. An example is an α-helix with PBE functional. This type of metallicity can be accurately treated in FMO/LCMO.

Diffuse functions can have a horrible convergence. To use them in FMO, special methods should be used (see Section 5.2).

4.10.1 *Charge Instability*

FMO can face charge instability problems, for which warnings are printed. These warnings should be taken very seriously and a major effort should be made to rectify them.

A charge instability occurs when HOMO of one fragment is above LUMO of another. In this case there is a tendency of the two electrons from the HOMO fragment to flow into the LUMO fragment (Figure 18). This is, however, a simplified view, because orbitals in a dimer are coupled, and they differ from MOs in the two monomers.

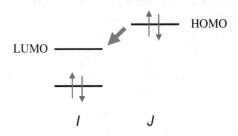

Figure 18.   Charge instability problem: LUMO of fragment *I* is lower than HOMO of fragment *J*.

A transfer of two electrons is bad for the accuracy of the MBE, which is based on the assumption that many-body corrections are small. This proclivity of two electrons to jump may not be realized, and the dimer calculation, despite the instability, may converge to the excited state, that resembles the combined density distribution in the two monomers. This is often the case if two fragments are far separated, whereas for short contacts the electron pair transfer perpetuated by an orbital overlap usually takes place. Using RSTRCT = .T. in $SCF may be helpful in getting a dimer state consistent with the two monomer states. For charge instability in dimers, SCF often has a bad convergence.

Trimers can have a likewise instability, which is even more complex than in dimers. A trimer with an interfragment separation close to the threshold RITRIM can fail to converge, in which case a somewhat radical measure may be to lower the threshold in an attempt to exclude the problematic trimers.

The most common cause of such instability is a zwitterionic charge distribution, because anions have high-lying HOMOs and cations low-lying LUMOs. A solvent can rectify the problem by preferentially stabilizing the occupied orbitals (including HOMO). Calculating zwitterionic systems in the gas phase is especially problematic with DFT and DFTB, but it is easier with their long-range corrected versions.

An example of charge instability warnings for a small protein with multiple charged residues in the gas phase is shown in Table 61. Most fragments involved in the report are charged: 3, 5, and 10 are anions, with a high HOMO. There is only one cation in this protein, fragment 1, which does not pose a problem. Interestingly, an instability is found for two neutral fragments, 6 (high HOMO) and 9 (low LUMO). Apparently, a strong electrostatic embedding in vacuum can induce such behavior in neutral residues.

The results in Table 61 indicate a catastrophe: fragments 3, 5, 6, and 10 have negative ionization potentials (IP). This means that these fragments can be stabilized by getting rid of an electron. Normally, IPs are positive. In this system, electron affinities (EA) are positive,

Table 61. Charge instability summary for a protein calculated in vacuum.

Fragment pair affected			
Charge instability detected: 3- >	2 IP =	−0.8697 EA =	0.6948 (eV)
Charge instability detected: 3- >	9 IP =	−0.8697 EA =	0.3678 (eV)
Charge instability detected: 5- >	2 IP =	−2.3328 EA =	0.6948 (eV)
Charge instability detected: 5- >	9 IP =	−2.3328 EA =	0.3678 (eV)
Charge instability detected: 6- >	9 IP =	−0.3700 EA =	0.3678 (eV)
Charge instability detected: 10- >	2 IP =	−0.9684 EA =	0.6948 (eV)
Charge instability detected: 10- >	9 IP =	−0.9684 EA =	0.3678 (eV)

which is normal (acquiring an extra electron is destabilizing). However, in stable calculations EAs are usually larger (because LUMOs are higher). IP and EA are estimated according to the Koopmans' theorem.

Salt bridges (two oppositely charged nearby fragments) may be merged into one fragment eliminating the instability (although in solution salt bridges may be quite stable).

A charge instability may arise due to user mistakes in preparing an initial structure, for example, with a wrong protonation, charge assignment, or a poor geometry.

The advice to deal with a charge instability is to examine the structure closely, and refine it as well as possible. One has to critically consider the protonation, QM level (the DFT functional or DFTB parameters), and basis set. It may be useful to merge problematic pairs of fragments with opposite charges.

4.10.2 *Basic HF/DFT Convergers*

There are two main choices for HF convergers, DIIS or SOSCF, selected by their names set to .T. (true) or .F. (false) in $SCF. The strength of DIIS is that it can tame shrew fragments, so it is often employed in UHF and DFT; its weakness is that it has a hard time finishing off, that is, to reach the desirable level of SCF convergence, so a calculation almost but not fully converges. The strength of SOSCF is that if it does not break in the beginning (which it can because it uses an approximation of the orbital Hessian), it will usually converge rather quickly. SOSCF is often the best choice for HF, and DIIS for DFT.

There are other points of difference. SOSCF does orbital optimizations by rotating the MO matrix, so that it does not need to diagonalize the Fock matrix, except in the first and the last SCF iterations. Matrix diagonalizations can be a major efficiency bottleneck of SCF on massively parallel computers. Moreover, SOSCF is better parallelized than DIIS, and, overall, the cost per iteration in SOSCF may be lower than in DIIS.

For a bad initial guess, a converger other than SOSCF or DIIS can be employed in the few initial iterations. For it, adjusting SOGTOL (ETHRSH) in $SCF may be useful to delay engaging SOSCF (DIIS).

4.10.3 *Fine-tuning Convergers*

There are several "levers" that can be applied to help either DIIS or SOSCF. One problem sometimes arises with direct SCF, that the calculation almost but not fully converges to the desired level, because of numerical errors accumulated due to the FDIFF option (used for updating the Fock matrix via density differences). In this case, the option DIRTHR in $SCF may help (for instance, DIRTHR = 1e-7), which recalculates the Fock matrix from scratch once the convergence reaches the specified threshold.

LOCOPT = .T. is helpful for some difficult to converge fragments. It prevents an internal reset of the values of SHIFT and DAMP which can happen with SWOFF in $DFT or SWDIIS in $SCF. RESET = .T. in UHF resets convergers if the energy rises in SCF.

For a small HOMO-LUMO gap in a fragment, i.e., for a fragment metallicity, one can attempt to use $SCF SHIFT = .T. (level shifting) and/or DAMP = .T. (Fock matrix damping).

4.10.4 *DFTB Convergers*

In DFTB, convergers are known as charge mixers. The SCC embedding in DFTB is based on Mulliken atomic charges, obtained from the electron density.

DFTB has its own set of mixers, of which the Broyden method (ITYPMX = 0 in $DFTB) is the default except that for LC-DFTB, ITYPMX = −1 is used to engage common SCF convergers defined in $SCF (e.g., SOSCF).

DFTB overall has a very good convergence; if something does not converge, it is usually either due to a bad structure or a wrong approach (such as trying to compute a zwitterionic system in gas phase without an LC method). Altering details of DFTB mixers rarely helps. Electronic temperature cannot be used in FMO.

4.10.5 *MCSCF Convergers*

There are two main choices, FULLNR and SOSCF. FULLNR somewhat resembles DIIS for HF, that is, it is more robust than SOSCF but sometimes fails to reach the required convergence level while getting very close to it. SOSCF, unless it breaks in the first initial iterations, often converges quite well.

The convergence of MCSCF is strongly governed by the quality of the initial guess. It is common to prepare initial orbitals from some preliminary run and read them in as a restart (Section 5.5.3). This can be done for both monomers and dimers.

4.10.6 *Alternating QM Methods*

Bad convergence in DFT can often be solved with the COROFF option in which DFT is replaced with HF after several iterations in the monomer loop (for the sake of not offending sensitivities, one can say that the functional in DFT is changed to the Slater with 100% HF exchange). COROFF only affects monomers, whereas for turning off DFT in some initial SCF iterations of dimers and trimers, SWOFF in $DFT can be used.

In MCSCF, a full RHF calculation can be done for a dimer before doing MCSCF, accomplished with NGUESS = 18 in $FMOPRP. Owing to an automatic matching of active orbitals, this approach can work quite well. The matching uses overlaps to find orbitals in the RHF dimer that most closely resemble monomer MCSCF orbitals to preserve the active space.

If 2 is added to NOPFRG of the MCSCF fragment (for example, NOPFRG(16) = 2 for fragment 16), then, after the preliminary RHF converges, modified virtual orbitals are calculated for cation + 6 as in MVOQ = 6 of $SCF (the value of 6 cannot be changed).

4.10.7 *Mixing Convergers*

In case of a bad convergence, it is possible to mix convergers and take advantage of their peculiarities. There are several ways of doing it.

In the first approach, NCVSCF in $FMOPRP is used. For example, by setting NCVSCF = 3, the first three monomer SCF iterations are performed with the converger set in $SCF, typically, with DIIS. After that, the converger is reversed in the pair DIIS/SOSCF: if it was DIIS, then SOSCF is used and vice versa. SWDIIS in $SCF can be set to some value like 1e-3 in order to use DIIS for some initial iterations of dimers and trimers and switch to SOSCF after that.

In the second approach, which can be combined with the first, a different converger can be explicitly specified as MCONV in $FMOPRP for each step in FMO (Table 1) by defining a packed value according to Table 62. For example, MCONV(1) = 17,65 sets up DIIS for monomers (16 + 1) and SOSCF (64 + 1) for dimers, both without direct SCF. To use direct SCF with FDIFF, add 768 (for example, for SOSCF with direct SCF, use 65 + 768 = 833).

It is possible to execute an even finer level of control and select a desired SCF converger for specific fragments, using MCONFG in $FMOPRP. To specify SOSCF + EXTRAP + DAMP + SHIFT with direct SCF for fragment 5, use MCONFG(5) = 839 (1 + 2 + 4 + 64 + 768), and such options can be set for multiple fragments, as in $FMOPRP MCONFG(5) = 839 MCONFG(12) = 17 MCONFG(15) = 65 $END.

SCF convergers can also be selected for up to 10 dimers and tri-mers. This is done by defining quadruplets I,J,K,M starting from MCONFG(N + 1), where N is the number of fragments. Here, I, J, and

Table 62. Packing of option for SCF convergers (more values can be found in docs-input.txt).

Value	$SCF	$MCSCF[a]
1	EXTRAP	
2	DAMP	
4	SHIFT	
16	DIIS	
64	SOSCF	
128	LOCOPT	
256	FDIFF	
512	DIRSCF	
1024		FOCAS
2048		SOSCF

[a]FULLNR is selected by not setting FOCAS and SOSCF options.

K select an n-mer (these three indices should be sorted with larger values coming first), and M is the packed converger option.

It is possible to repeat SCF or a fragment with an alternative converger, if SCF diverged, accomplished with 256 added to NGUESS in $FMOPRP. If 2048 is added to 256 (implemented only for MODORB = 0), the alternative converger will not use unconverged orbitals as its initial guess.

4.10.8 *Convergence Criteria*

In Table 63, a summary of various criteria is presented as pertinent to iterative approaches. OPTTOL and NSTEP can be changed in geometry optimizations. For most other options, the philosophy is that if something does not converge in the default way, it is likely that there is a bigger issue of which divergence is a symptom, so rather than fighting the symptom, the user should seek the real cause.

Table 63. Covergence thresholds and maximum iteration limits.[a]

Description	Calculation	Criterion	Limit	Input group
SCF (SCC)	HF, DFT, DFTB	CONV = 1e-7	MAXIT = 30	$CONTRL
MCSCF	MCSCF	CONV = 1e-6	MAXIT = 30	$MCSCF
Monomer loop	general	CONV = 1e-7	MAXIT = 30	$FMOPRP
PCM loop	PCM[m], $m > 1$	CNVPCM = 1e-7	NPCMIT = 30	$FMOPRP
Solvent charges in Eq. (22)	PCM, SMD	THRES = 1e-8	MXITR1 = 50 MXITR2 = 50	$PCMITR
CI diagonalizer	MCSCF	CVGTOL = 1e-6	ITERMX = 100	$DET
Excited states	TDDFT, TDHF	CNVTOL = 1e-7		$TDDFT
CP-DFTB	DFTB	CPCONV = 1e-6	MXCPIT = 50	$DFTB
CCSD solver	CCSD	ICONV = 8[b]	MAXCC = 30	$CCINP
Geometry optimization	OPTIMIZE SADPOINT	OPTTOL = 1e-4	NSTEP = 200	$STATPT
Geometry optimization	OPTFMO	OPTTOL = 1e-4	NSTEP = 200	$OPTFMO
Surface crossing	MEX	TGMAX = 5e-4 TGRMS = 3e-4	NSTEP = 50	$MEX
Bond constraints	RATTLE in MD	RATTOL = 1e-6	IRATIT = 2000	$MD

[a]Typical default values are shown, but they may depend on other options.
[b]The value of 8 means 10^{-8}.

The options in Table 63 are the most common. There are many other less conspicuous thresholds, which are not listed. Some iterative loops, most importantly, in SCZV and CPHF, have no input options to affect convergence. It is worth mentioning that TIMLIM in $CONTRL defines the time limit (in minutes), and is 1 year by default.

4.11 Fragmentation

For a covalent bond between fragments, a special treatment (called detachment) is needed. There are two types of it, hybrid projection operator (HOP) and adaptive frozen orbital (AFO). It is for the user to choose one in an FMO calculation. Typically, HOP is used for chains (proteins, DNA, organic polymers, etc.), so that on the border of two fragments there is a single covalent bond, and AFO is used for inorganic systems with multiple bonds to be detached between a pair of fragments. This is a somewhat simplified view, as there is nothing in each method restricting its application; it is rather a practical advice based on the accuracy and peculiarities of each method.

The main difference between them is in freezing (AFO) or relaxing (HOP) the MO that describes the detached bond. The second practically important difference is that hybrid orbitals used in the boundary treatment should be prepared in advance for HOP, whereas for AFO they are made on the fly.

There is no clear winner comparing the accuracy of AFO and HOP in general, although for inorganic systems with multiple detached bonds per boundary the accuracy of AFO tends to be higher, whereas for polar systems such as proteins HOP may be more accurate. The accuracy of MP2 for AFO seems to be systematically worse than for HOP. AFO, on the other hand, may be more accurate for HF with large basis sets.

Both methods share the particle division (Figure 19). A covalent bond on the boundary is between a bond-detached atom (BDA) and a bond-attached atom (BAA). BDA and BAA contribute one electron each toward making the bond, and these two electrons are assigned to the BAA fragment. To keep fragments neutral, one proton is also reassigned from the BDA fragment to the BAA fragment. The proton reassignment is formal and does not create any artefacts, because a fragment calculation is done in the embedding potential, and

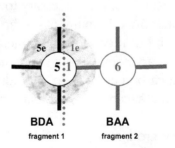

Figure 19. Fragmentation of a C–C bond (red dashed line). The carbon BDA has 6 protons and 6 electrons (e), which are formally divided between two fragments (black and blue) as 5:1. However, each fragment is computed in an embedding potential, which puts back all particles.

any proton reassigned to another fragment is added back as the embedding potential. The electron reassignment is the source of some artefacts, for example, the pair interaction for a detached bond is large (Section 3.4).

One important aspect of bond detachment is that a BDA is present in two fragments. It is on the border, and its electrons and basis functions are used in two different fragments. Its basis functions are redundant, and the redundancy is eliminated for both HOP and AFO.

A bond detachment in FMO affects two electrons, which means that it is a single bond that is detached. This is the current limitation of FMO in GAMESS. A stretching example of a bond detachment is a B–N bond in a B_3N_3 ring, which is isoelectronic to a benzene ring C_6. B_3N_3 rings occur in white graphene and its various derivatives, such as BN nanotubes and ribbons.

Another stretching example is a branching BDA, that is, a BDA that is involved in more than one detached bond. In neopentane $C(CH_3)_4$, the middle carbon is an example of a branching BDA, if more than one C–C bond is detached, with the central atom being BDA in each. Such a division is not recommended due to considerations of accuracy.

Fragmentations, where an atom is a BDA in one bond and BAA in another, are discouraged. Detached bonds should be as far as possible from each other. For ligands and other organic molecules, there is some freedom of fragmentation. Dividing a ligand into multiple pieces, if done at all, should be attempted with reluctant care.

For a bond to be detached, it is necessary to decide which atom is to be BDA and which BAA. In many cases the difference in properties is not substantial, but it can also be large. The general advice is to choose the BDA as an atom without lone pairs, and with a small atomic charge (e.g., a carbon atom in sp^3 hybridization). For detaching a C–O bond (as in a saccharide), the carbon atom may be preferred as the BDA.

Splitting functional groups, such as OH or COOH as fragments, is in general discouraged, because it leads to a poor overall accuracy and the results are exceedingly affected by the artefacts of fragmentation. For defining properties of functional groups, PA can be used.

In designing a new fragmentation, a good understanding of the electron distribution is needed, and an inconsiderate detachment of bonds may easily lead to disastrous results. A good starting point is to do a regular QM calculation of a model system with the bond to be detached still intact, and look at the electron density distribution and localized MOs in detail, paying attention to whether multiple atoms may be involved in the bond and what the bond order is.

Whenever a novel fragmentation is attempted, and the user is not sure if the fragmentation is suitable or not, it is advised to do an unfragmented calculation and compare the total energy with that in FMO. Because FMOn is exact for a system divided into n fragments, doing FMO2 for a ligand divided into 2 fragments and comparing it to an unfragmented calculation is of no use. For a comparison to be useful, there should be at least 3 fragments in FMO2.

A useful measure to indicate that a fragmentation is dubious is the interfragment CT $\Delta Q_{I \to J}$. The larger the values, the more alarming they are. A hydrogen bond has typical values of 0.02–0.05 e, which are fine. The values of 0.1 are seen in proteins for charged residues and may be tractable. Values larger than 0.2 should be considered alarming, and a refragmentation may be attempted.

One troublesome example is a salt bridge in a protein with two residues forming a strong bond. If they get into close contact, merging them into one fragment may be considered.

Below, fragmentation of various kinds of systems is discussed according to their type. If there are several types in one system, for example, a protein in explicit water, then the fragmentation for each type (protein and water) is done separately and the results are combined (programs for making input files can handle such combination cases automatically).

4.11.1 *Hybrid Orbital Projection (HOP)*

HOP is a simpler approach than AFO because polarization is not restricted. An operator (HOP) is added to the Hamiltonian to divide basis functions of each BDA between two fragments. Consider a carbon atom, which has 1s core orbital and forms four sp^3 orbitals.

One sp^3 orbital that describes the bond to be detached should be retained by the BAA fragment. The other three sp^3 and one 1s orbital stay with the BDA fragment (see Figure 19, where 4 σ bonds (shown as bars) of the BDA are divided as 3:1 between the two fragments, and 1s core orbital (not shown) is assigned to the left fragment, so overall, occupied orbitals are divided as 4:1).

To use a HOP treatment, one has to prepare occupied orbitals for each unique BDA. In a protein, C–C bonds are detached and BDAs are always carbon atoms. In this case, a single set of hybrid orbitals is needed for all carbon BDAs. The hybrid orbitals are basis set dependent, so for each basis set, if there are several, a separate set of hybrid orbitals is needed. If both C–C and Si–Si bonds are detached in one calculation, then two sets of hybrid orbitals are needed, one for C and another for Si. No orbitals are needed for a BAA, because AOs for BAAs are not redundant (each BAA is placed in 1 fragment only).

In principle, different hybridizations may call for separate sets of orbitals. Consider a system with a $=C'|-C-$ bond, and a $-C''|-C$ bond, detached (|) at C' and C''. One could generate a separate sp^2 set of hybrid orbitals but in the author's experience it does not improve the accuracy, and the same set of sp^3 orbitals can be used for detaching σ-bonds of both sp^2 C' and sp^3 C''.

Hybrid orbitals are provided for sp^3 carbon BDAs in tools/fmo/ HMO: HMOs.txt (Pople basis sets like 6-31G*), HMOS.spher.txt (Dunning basis sets like cc-pVDZ), and HMO.DFTB.txt (DFTB). If the desired basis set is not there, or a BDA is not carbon, then the user has to make hybrid orbitals.

The provided HMOs are generated with HF (or DFTB). Generating HMOs for specific methods (e.g., DFT) can be done but it does not seem to improve the accuracy.

To generate HMOs (Table 64), one can use makeHMO.inp provided in tools/fmo/HMO/. First, it is necessary to choose a model system. For example, for an sp^3 carbon, CH_4 works well. For sulfur, one can try SH_2. Although possible, it does not seem to be beneficial to use larger model systems, such as CH_3–CH_3 for an sp^3 carbon. As for geometry, a natural choice is to optimize it.

Table 64. How to make hybrid orbitals for HOP, exemplified for sulfur.

Step	Description
1	Choose a model system (SH_2), possibly optimize its geometry.
2	Rotate the model system so that the model bond (any S–H) is along the z-axis.
3	Run a single point calculation to localize the orbitals.
4	Extract localized (= hybrid) orbitals from the output, and process the matrix.
5	(Optional: normalize hybrid orbitals.)
6	Put the orbitals in the format of $FMOHYB.

It is necessary to reorient the molecule so that one bond of the desired type is along the z-axis, because hybrid orbitals provided in $FMOHYB have to be rotated inside the FMO code so that they match the actual bond vector. The convention that must be obeyed is that the orbitals are generated for the model bond aligned along the z-axis (parallel to it is good enough).

GAMESS expects a certain orientation of atoms for each point group. CH_4 has T_d symmetry if atoms are placed as GAMESS wants them to be, but if one bond is aligned along the z-axis, the symmetry becomes C_{3v} (as far as GAMESS can use it). Although using symmetry as much as possible is encouraged, in case of doubt, one can always use C_1 (no symmetry). For a reoriented SH_2, no symmetry can be (easily) used.

If necessary, a molecule can be rotated so that the desired bond is properly aligned along the z-axis. The rotation axis **n** can be defined as

$$n = \frac{z \times b}{|z \times b|} \tag{81}$$

where $z = (0,0,1)$, and $b = R_H - R_{BDA}$ is the vector pointing from BDA (sulfur) to the hydrogen representing BAA, defined using their coordinates R (if **b** is collinear to **z**, and $|z \times b| = 0$, then the molecule is already oriented and there is no need to do this rotation).

The molecule is to be rotated around **n** by the angle ω,

$$\omega = \arccos\left(\frac{\mathbf{z} \cdot \mathbf{b}}{zb}\right), \tag{82}$$

accomplished with the $(\mathbf{n}, \omega)$ representation of the SO(3) group,

$$\mathbf{U} = \cos\omega\,\mathbf{I} + (1 - \cos\omega)\begin{bmatrix} n_x^2 & n_x n_y & n_x n_z \\ n_y n_x & n_y^2 & n_y n_z \\ n_z n_x & n_z n_y & n_z^2 \end{bmatrix} + \sin\omega\begin{bmatrix} 0 & -n_z & n_y \\ n_z & 0 & -n_x \\ -n_y & n_x & 0 \end{bmatrix} \tag{83}$$

where $\mathbf{I}$ is a unit matrix 3×3.

The coordinates of each atom $\mathbf{R}_A$ can be shifted to the BDA and rotated as

$$\mathbf{R}'_A = \mathbf{U}(\mathbf{R}_A - \mathbf{R}_{\text{BDA}}) \tag{84}$$

A critical mind would compare the energy of the reoriented molecule to the original energy, making sure they are identical.

Table 65.   Sample input for computing localized orbitals in SH_2.

Input	Comments
$CONTRL SCFTYP = RHF LOCAL = RUEDNBRG $END	
$SYSTEM TIMLIM = 9999 MEMORY = 10000000 $END	
$BASIS GBASIS = STO NGAUSS = 3 $END	
$DATA	
SH2 OPTIMIZED AND ROTATED	
C1	
S 16 0.0000000000 0.0000000000 0.0000000000	BDA (S)
H 1 0.0000000000 0.0000000000 1.3550624824	This S–H bond is
H 1 0.0000000000 1.3523960511 −0.0849661828	along the z-axis.
$END	

Next, a single point calculation is executed to localize molecular orbitals. The provided file tools/fmo/HMO/makeHMO.inp combines a geometry optimization with making localized orbitals in one run. This is possible for symmetries that force the bond to stay in the proper direction during optimization, but if a manual reorientation is needed, two separate runs are necessary. For an orbital localization, use LOCAL = POP in $CONTRL for DFTB and LOCAL = RUEDNBRG for other methods. A sample input file for computing localized orbitals is shown in Table 65.

Table 66. Localized orbitals in SH$_2$ (RHF/STO-3G).

EDMISTON-RUEDENBERG ENERGY LOCALIZED ORBITALS

				1	2	3	4	5
1	S	1	S	**-0.994204**	-0.368151	0.000000	-0.006562	0.000000
2	S	1	S	-0.016970	**1.049811**	0.000000	0.018971	0.000000
3	S	1	X	0.000000	0.000000	0.000000	0.000000	**-0.988863**
4	S	1	Y	-0.000135	0.010821	**0.675471**	**-0.719298**	0.000000
5	S	1	Z	-0.000126	0.010163	**-0.719240**	**-0.675526**	0.000000
6	S	1	S	0.002350	0.049935	0.000000	-0.011276	0.000000
7	S	1	X	0.000000	0.000000	0.000000	0.000000	-0.044401
8	S	1	Y	0.000316	0.006489	0.037642	-0.039453	0.000000
9	S	1	Z	0.000296	0.006094	-0.040081	-0.037052	0.000000
10	H	2	S	-0.000636	-0.010931	0.011571	0.012074	0.000000
11	H	3	S	-0.000636	-0.010931	-0.011571	0.012074	0.000000

				6[a]	7	8	9
1	S	1	S	0.027208	0.027208	-0.052277	-0.052277
2	S	1	S	-0.097354	-0.097355	0.172486	0.172486
3	S	1	X	0.000000	0.000000	-0.197348	0.197348
4	S	1	Y	0.003472	-0.174476	-0.072130	-0.072130
5	S	1	Z	-0.175039	0.014440	-0.067740	-0.067740
6	S	1	S	0.225675	0.225676	**-0.670717**	**-0.670717**
7	S	1	X	-0.000001	-0.000001	**0.725869**	**-0.725870**
8	S	1	Y	-0.002289	**0.559280**	0.252592	0.252591
9	S	1	Z	0.560527	-0.037431	0.237221	0.237220
10	H	2	S	0.545133	-0.081140	0.099339	0.099338
11	H	3	S	-0.081140	**0.545132**	0.099339	0.099338

[a]Orbitals 1–5 are essentially atomic orbitals of S. Orbital 6 is the $\sigma(S–H_1)$ orbital for the special bond (along the z-axis). It has large valence $p_z(S)$ and $s(H_1)$ coefficients. Orbital 7 is another $\sigma(S–H_2)$, and orbitals 8–9 are lone pairs.

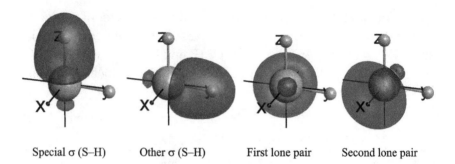

Special σ (S–H)   Other σ (S–H)   First lone pair   Second lone pair

Figure 20. Valence orbitals in SH_2. One hydrogen atom H_1 is along the z-axis.

The obtained localized orbitals are shown in Table 66. They are printed as columns, that is, as $C_{\mu i}$, where i is a localized MO, and μ is an AO. First, one has to locate all MOs i relevant for the BDA. For sulfur, these are: 1s, 2s, 2p (5 core orbitals) and 4 sp³-like orbitals (whether or not they are properly to be called sp³ is a technical issue not delved into here). The total number of occupied orbitals for S is thus 9, so one has to find 9 orbitals among localized MOs. Among those, core orbitals are easy to spot because they are essentially atomic orbitals of S, with small coefficients in the valence set of primitives. To pick the right orbitals, it may be convenient to use MacMolPlt to visualize them (Figure 20). Two sp³ orbitals are lone pairs, and another two are σ(S–H) bonds, with hydrogen AOs mixing in.

From the matrix $C_{\mu i}$, one has to cut the block for $\mu \in$ S, with i limited to localized MOs of interest (nine of them). LCAO coefficients for hydrogens are omitted. If the basis set is STO-3G, the matrix thus obtained should be 9 (AO) × 9 (MO). Then the matrix is transposed, from $C_{\mu i}$ to $C_{i\mu}$, so that now LMOs are rows.

Next, the order of the localized (= hybrid) orbitals should be changed, by moving LMO #6, which describes the special σ bond (Figure 20), with a large valence p_z coefficient (0.5605), to be the first orbital in the list. This is done by changing the order of rows. The order of the other 8 orbitals is irrelevant. If one strives for perfection, they can be reordered just for the sake of beauty rather than any

Table 67. Transposed matrix of LMOs for S in SH$_2$ prior to reordering (LMOs are rows).[a]

-0.9942	-0.0170	0.0000	-0.0001	-0.0001	0.0024	0.0000	0.0003	0.0003
-0.3682	1.0498	0.0000	0.0108	0.0102	0.0499	0.0000	0.0065	0.0061
0.0000	0.0000	0.0000	0.6755	-0.7192	0.0000	0.0000	0.0376	-0.0401
-0.0066	0.0190	0.0000	-0.7193	-0.6755	-0.0113	0.0000	-0.0395	-0.0371
0.0000	0.0000	-0.9889	0.0000	0.0000	0.0000	-0.0444	0.0000	0.0000
0.0272	**-0.0974**	**0.0000**	**0.0035**	**-0.1750**	**0.2257**	**0.0000**	**-0.0023**	**0.5605**
0.0272	-0.0974	0.0000	-0.1745	0.0144	0.2257	0.0000	0.5593	-0.0374
-0.0523	0.1725	-0.1973	-0.0721	-0.0677	-0.6707	0.7259	0.2526	0.2372
-0.0523	0.1725	0.1973	-0.0721	-0.0677	-0.6707	-0.7259	0.2526	0.2372

[a]Some trailing significant figures are truncated. The special orbital is shown in bold.

practical use, to be in the order of 1s, 2s, 2p, and the remaining three sp^3 orbitals (which they naturally tend to be anyway).

The next step of normalizing the hybrid orbitals is optional. After removing the coefficients of hydrogens, LMOs are no longer normalized: $C_i^t SC_i$ is not equal to 1, where S is the overlap matrix. The normalization has a negligible effect on results and it can be skipped.

Finally, the matrix $C_{i\mu}$ should be rewritten in the format of $FMOHYB (see Section 2.6.5). This is done by devising a name (not more than 8 characters), such as S-STO3GU, followed by the number of AOs for sulfur (9) and then the number of hybrid orbitals (9). Then for the first (special) orbital, a designation "1 0" is added, which means that this orbital is projected out from the fragment containing BDA, and for all 8 other orbitals "0 1" is used, which means that they are projected from the fragment containing BAA. After that, a hybrid orbital i is placed as a row of coefficients $C_{i\mu}$ for all AOs μ. There should be 9 such rows, corresponding to 9 values that i can take. There is a limit of 80 characters per line, so if necessary, the list of coefficients can be split into multiple lines. The result is shown in Table 68.

Here, it should be noted that to use these orbitals in FMO, MAXCAO in $FMO should be increased. It defines the maximum number of occupied HMOs, and the default value is 5 (as in C).

Table 68.   HMO set for S (unnormalized and formatted for 80 characters per line).

```
$FMOHYB
S-STO3GU 9 9
1   0   0.027208    -0.097354    0.000000     0.003472    -0.175039
        0.225675    -0.000001   -0.002289     0.560527
0   1  -0.994204    -0.016970    0.000000    -0.000135    -0.000126
        0.002350     0.000000    0.000316     0.000296
0   1  -0.368151     1.049811    0.000000     0.010821     0.010163
        0.049935     0.000000    0.006489     0.006094
0   1   0.000000     0.000000    0.000000     0.675471    -0.719240
        0.000000     0.000000    0.037642    -0.040081
0   1  -0.006562     0.018971    0.000000    -0.719298    -0.675526
       -0.011276     0.000000   -0.039453    -0.037052
0   1   0.000000     0.000000   -0.988863     0.000000     0.000000
        0.000000    -0.044401    0.000000     0.000000
0   1   0.027208    -0.097355    0.000000    -0.174476     0.014440
        0.225676    -0.000001    0.559280    -0.037431
0   1  -0.052277     0.172486   -0.197348    -0.072130    -0.067740
       -0.670717     0.725869    0.252592     0.237221
0   1  -0.052277     0.172486    0.197348    -0.072130    -0.067740
       -0.670717    -0.725870    0.252591     0.237220
$END
```

For S, MAXCAO should be increased to 9. The number of AOs in S (happens to be 9 as well, because STO-3G is a minimum basis set) is determined automatically and there is no need to set it.

Now, although sulfur and SH_2 were chosen as an example for generating hybrid orbitals, it is hard to find a molecular system where they may be useful. One is tempted to use these orbitals to cut sulfur bridges (S–S bonds) in proteins, but the accuracy for such fragmentation seems to be poor. This may be related to the above advice of avoiding fragmentation at atoms with lone pairs. It can be verified by devising a small polypeptide with a sulfur bridge, detaching the S–S bond and checking the total FMO energy against the energy without fragmentation. The only case when a S–S detachment seems to work reasonably well is DFTB.

How does using a core potential affect HOP? Because a core potential removes some core electrons, HMOs are to be provided only for valence electrons. No orbitals are pregenerated for a core potential, and the user has to prepare them.

4.11.2 *Adaptive Frozen Orbitals (AFO)*

The AFO scheme is much more complicated than HOP inherently, although from the user point of view it may delusively appear simpler. For each detached bond, a model system is constructed automatically by following these steps: (a) BDA and BAA of the bond are added, (b) all atoms from the real system within a threshold radius from BDA and BAA are included, and (c) hydrogen caps are added. The algorithm to add hydrogen caps is very simple: it finds any bonds that (a) + (b) atoms form to other atoms in the real system, and replaces these other atoms with hydrogens, adjusting the bond length. As shown in Figure 21, the algorithm can fail to produce an appropriate model system. If an appropriate model system is not generated automatically, it can be manually designed (Section 2.6.6).

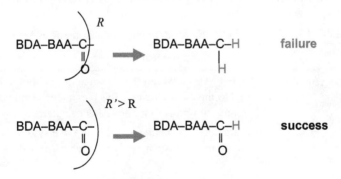

Figure 21. Inappropriate (upper panel) and appropriate (lower panel) model system construction for AFO, as determined by the distance threshold *R* (*R'*). Automatically added hydrogen caps are in red.

If the charge of a model system is not zero, which can happen if it includes a charged functional group, then the charge should be specified in $FMOHYB (Section 2.6.5).

For a model system, an SCF calculation (RHF, DFT, or DFTB) is performed with the basis set of the current layer. Then, molecular orbitals are localized with LOCAL = POP (DFTB) or RUEDNBRG (other methods). From these localized orbitals a matrix $C_{i\mu}$ similarly to HOP is automatically extracted. A big difference is that for HOP, $\mu \in$ BDA (one atom only), whereas for AFO, μ spans multiple atoms.

Border orbitals ($C_{i\mu}$) are used to remove the redundancy in the AOs of the BDA and to freeze the MO of the detached bond in the real system. The freezing step is absent in HOP. The idea behind it is that detached bonds in fragments may be deformed due to artefacts of fragmentation and it is advantageous to force them to have the right shape. The technical aspects of removing the redundancy and freezing the orbital are accomplished via a transformation of the Fock matrix, differently from the operator form used in HOP, but nearly equivalent as far as removing redundancy is concerned.

Conducting an AFO calculation may be slightly easier for the user, as there is no need to prepare hybrid orbitals. However, AFO calculations have various problems peculiar to them. Namely, the exact analytic gradient is available only for DFTB, and for other methods the gradient is approximate. There is no Hessian for AFO,

Table 69.   Options in $FMO for defining model systems in AFO.

Option[a]	Meaning
RAFO(1)	All atoms in the real system within this distance from either BDA or BAA are added to the model system. This parameter (denoted by R in Figure 21) must be non-zero for any AFO calculation.
RAFO(2)	Hydrogen caps are added by taking atoms in the real system, within RAFO(2) from the model system, and replacing them with hydrogens.
RAFO(3)	The set of AOs $\{\mu\}$ used to expand $C_{i\mu}$ is centered on atoms within RAFO(3) from either BDA or BAA.

[a]RAFO values are unitless, applied to relative distances as in Eq. (21).

PBC, UHF, ROHF, UDFT, and MCSCF and may not be used with AFO.

Two standard choices are suggested. In the large model, chosen with RAFO(1) = 1,1,1, all atoms covalently bound to BDA and BAA in the real system are included, and hybrid orbitals span AOs on all of these atoms. This model can be used for proteins. In the small model, chosen with RAFO(1) = 0.1,1,1, the smallest possible model is constructed, containing BDA, BAA and caps. This model is used for inorganic systems.

For large values of RAFO(1), it is likely that the automatic algorithm for adding caps will fail (Figure 21), which often happens for dense networks of bonds found in inorganic systems. On the other hand, the description of a detached bond may be better if more environment is included (i.e., a larger RAFO(1) may be more accurate).

It is possible to replace some automatically constructed model systems with manually prepared geometries, by placing them in $AFOMOD (Section 2.6.6). If multiple model systems should be replaced, they are put into one $AFOMOD group, one after another.

4.11.3 *BDA Corrections for PIEs*

Pair interaction energies (PIEs) between fragments connected by a covalent bond are on the order of 15 hartree for carbon BDAs. The bond detachment results in an exclusion of some components in the energies of monomers related to the electron division in the BDA, which are recovered in the dimer calculation. A large PIE for connected fragments includes an artificial contribution due to the fragmentation rather than a normal physical interaction.

Usually, a difference in two energies is of interest, for example, the difference in the energies of a transition state and reactants. When two large PIEs are subtracted, the artificial effects mostly cancel out, because they are local to the bond. For analyzing binding energies or reaction barriers, no BDA corrections are needed, but they may be useful for discussing interactions in a complex.

The idea of BDA corrections is that for a bond in question, minimal caps are added in a model system, for example, CH_3–CH_3 is used to represent C–C bonds. The model system is divided into two fragments across the BDA-BAA bond (resulting in two CH_3 fragments). Then the PIE in this minimalistic system is assumed to be the artificial interaction due to the bond fragmentation, which can be used in a real system, for example, for a C_α–C bond in a protein. A PIE in a real system can be corrected using the model PIE as

$$\Delta E_{IJ} \text{ (corrected)} = \Delta E_{IJ} \text{ (computed)} - \Delta E_{IJ} \text{ (model)} \qquad (85)$$

A computed C–C PIE, on the order of 9000 kcal/mol, is reduced to a value on the order of 100 kcal/mol for a corrected PIE. This value is roughly on par with non-covalent interactions.

To calculate BDA corrections, the following has to be done. Any FMO calculation writes out the values of detached bond lengths to DAT file as R0BDA. The user has to prepare a model system and set up an FMO calculation for two fragments (such as CH_3–CH_3 for a C–C bond). In the model system, the first atom should be BDA, and the second BAA. Moreover, both BDA and BAA should be placed on the z-axis (if necessary, the rotation in Eq. (83) may be applied).

In this BDA input file, an array of actual detached bond lengths R0BDA is added. The size of the array is set as N0BDA (this parameter is used only to generate BDA corrections, and it should not be set when computed BDA corrections are used for a real system). The program will shift one fragment so that the BDA-BAA bond distance is the same as in R0BDA. A sample of an input file for BDA corrections is shown in Table 70. If there are different types of bonds, for example, C–C and Si–Si, then two separate runs for BDA corrections should be made, and the results combined.

The resultant BDA corrections are written to DAT file as E0BDA. They can be pasted to an FMO input, for example, of PIEDA. In contrast to fragments, segments need no BDA corrections because atoms are not shared between segments. Currently, only gas-phase BDA corrections can be computed, and an add-on dispersion model may not be used in them.

Table 70. Sample input file for BDA corrections.

Input	Comments
⎵$FMO NFRAG = 2 INDAT(1) = 1,2,2,2,1,1,1,2 $END	CH_3–CH_3.
⎵$FMOPRP MODORB = 3 IPIEDA = 1	PIEDA/PL.
N0BDA = 2	BDA corrections computed for 2 bonds.
R0BDA(1) = 1.52459598,1.52790685	C–C bond lengths (Å) from a real
⎵$END	system.
⎵$BASIS GBASIS = STO NGAUSS = 3 $END	The basis set is STO-3G.
⎵$FMOXYZ	Coordinates of the model CH_3–CH_3.
C 6.0 0.000 0.000 1.436	BDA, on the *z*-axis.
C 6.0 0.000 0.000 −1.436	BAA, on the *z*-axis.
H 1.0 1.681 0.970 −2.122	All other atoms follow.
H 1.0 −1.681 0.970 −2.122	
H 1.0 0.000 1.941 2.122	
H 1.0 −1.681 −0.970 2.122	
H 1.0 1.681 −0.970 2.122	
H 1.0 0.000 −1.941 −2.122	
⎵$END	
⎵$FMOLMO	Hybrid orbitals
STO-3G 5 5	for STO-3G.
1 0 −0.117784 0.542251 0.000000 0.000000	
0.850774	
0 1 −0.117787 0.542269 0.802107 0.000000	
−0.283586	
0 1 −0.117787 0.542269 −0.401054 −0.694646	
−0.283586	
0 1 −0.117787 0.542269 −0.401054 0.694646	
−0.283586	
0 1 1.003621 −0.015003 0.000000 0.000000	
0.000000	
⎵$END	
⎵$FMOBND	
−1 2 STO-3G	−BDA BAA
⎵$END	
⎵$DATA	
CH3–CH3	
C1	
H 1	
C 6	
⎵$END	

4.11.4   *Molecular Clusters*

For molecular clusters, such as nanodroplets of a liquid, the natural division is 1 molecule per fragment. To increase the accuracy, 2 molecules may be combined into 1 fragment, although the only useful way of doing it is to pair up molecules that are geometrically close to each other, and that is not a trivial task for a large cluster.

A somewhat special example is ionic liquids, which are a mixture of ions of opposite charge. One could attempt to pair up ions of the opposite charge in one fragment, or assign an ion per fragment. Because of a large CT, for a better accuracy it may be desirable to use FMO3.

4.11.5   *Atomic Ions*

Metal cations, such as Na^+, Ca^{2+}, etc., and anions (Cl^-, etc.) pose an accuracy problem for fragmentation. A fragment in FMO is assigned an integer charge, such as +1, +2, or −1, but these are formal textbook charges, and the actual charges on ions can be quite different, due to a CT between ions and neighboring atoms, for example, solvent molecules. Although technically it is possible to set up a fragment consisting of just an ion, it may not yield sufficient accuracy in terms of the total properties.

A common practical solution is to merge an ion and the ligands with which it forms dative bonds into one large fragment. This usually works well in terms of the accuracy, but complicates analyses of properties of ions. An alternative solution is to use FMO3 while assigning an ion per fragment (in this case, the option to check for short interatomic contacts between fragments should be disabled with PRTDST(3) = 0.4 in $FMOPRP).

In DFTB, it is possible to perform partition analysis using segments with fractional charges (Section 3.5.1), an attractive option for atomic ions. Damping of metal-hydrogen electrostatics in DFTB may be a problem (see Section 4.1.3), so it can be disabled.

As a curiosity, some cations, if treated as separate fragments, have no electrons, for example, Na^+ in DFTB or Zn^{2+} with SBKJC ECP (Section 4.3.1). DFTB has been well tested to treat such electron-less fragments, and to some extent non-DFTB methods may also be used, but perhaps not in every kind of run, because QM methods may find the lack of electrons too amusing.

4.11.6 *Polymers (Including Polypeptides, Polysaccharides, and Polynucleotides)*

For proteins, DNA, and polycarbohydrates, there is a well-tested fragmentation pattern that can be applied automatically, for example, using Facio.[13] As a result, for proteins including enzymes, residue fragments differ from conventional residues by a CO group, because the standard fragmentation is at C_α atoms in C_α–C bonds, as necessitated by accuracy considerations. Nucleic acids (DNA and RNA) can be split as 1 nucleotide per fragment (also shifted by a phosphate group compared to the conventional definition) and bases may be split. For polysaccharides, such as cellulose or heparin, there is some freedom where to fragment.

Sulfur bridges between two cysteine residues are not fragmented. Two cysteins are merged into one fragment, creating a disparity between fragments and residues, including their numbering. Strong salt bridges may be manually merged into 1 fragment because of a large CT. For some analyses (PA and PAVE), it is possible to define segments as conventional residues, facilitating the interpretation of results.

Other organic and inorganic polymers may be divided into fragments, with a possible exception of polymers with conjugation of double bonds that leaves no suitable place for fragmentation.

4.11.7 *The Junction Rule*

The junction rule[7] states that to describe a system with a junction of order n, at least FMOn should be used for good accuracy in the absolute energy (for relative energies the order may be decreasable). In a junction of order n, a boundary MO of a fragment has a substantial overlap with boundary orbitals of $n-1$ other fragments. Some examples are given in Figure 22.

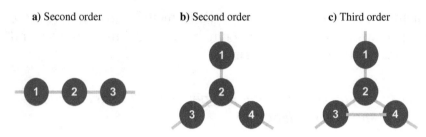

Figure 22.   Illustration of junctions between fragments (filled circles) connected by covalent bonds (blue lines).

In Figure 22a, consider fragment 2. There are boundary MOs of fragment 2 that overlap with boundary orbitals of fragment 3, because there is a detached bond between them, and BDA orbitals are shared between the two fragments. Therefore, considering the right side, the junction order n is 2 ($n-1 = 1$). Likewise, on the left side there is one fragment 1. Here, orbitals describing detached bonds at the 2-1 and 3-2 boundaries are well separated, with a small overlap between them.

In Figure 22b, there are also likewise pairs. Although there is a branching fragment 2, it does not increase the junction order, if fragments 1, 3, and 4 are sufficiently separated, and boundary MOs in each pair of fragments have a negligible overlap with boundary MOs in other fragments.

In Figure 22c, consider fragment 3. It has a detached bond with 2 and another detached bond with 4. The two detached bonds are geometrically close such that a boundary orbital of 3 has a substantial overlap with 2 fragments, 2 and 4, so the order of this junction is 3.

4.11.8 *Inorganic Systems (Nanomaterials)*

The junction rule provides the required order of FMO and effectively warns against such fragmentations for which a junction order is high. The published examples of fragmentations[7] may be studied. If there are many detached bonds between fragments, the parameter MAXBND in $FMO may have to be increased (set to the number of detached bonds plus 1).

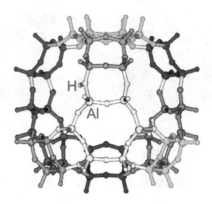

Figure 23. Fragmentation of a finite cluster of faujasite zeolite. The Al site and acidic H give rise to catalytic activity.

Fragmenting a solid chunk of a covalent crystal does not seem to be practical, because the order of junctions is too high. However, when the molecular system has some sparsity (pores) or a reduced dimensionality (oblong nanowires), then a fragmentation is possible.

A faujasite zeolite cluster can be divided into 10 fragments as shown in Figure 23, which corresponds to junctions of the second order (the pattern in Figure 22b). However, guest molecules tend to bind to a border between fragments, so for a better accuracy the two fragments in the active site can be merged (Section 6.6). This shows that one has to be attentive to the fragmentation throughout the study as the fragmentation chosen for a standalone system may not be the best in a complex.

The silicon nanowire was sliced as a piece of sausage into 6 fragments (Figure 24), with 9 bonds detached per fragment pair (Section 6.6). Because of the symmetry, the total dipole moment of the nanowire should be 0. To obtain an embedding mimicking this property, the zigzag arrangement of BDA-BAA pairs can be used, with their mirror assignment in the left and right parts (to cancel out the artefacts of fragmentation). The order of junction is 2 (as in Figure 22a). Facio[13] provides a convenient interface in which the bonds to be detached can be clicked on the screen.

Figure 24.   Division of a silicon nanowire into 6 fragments (the surface is saturated with hydrogens, appearing as goose bumps).

4.11.9 *Periodic Systems (Liquid and Solid State)*

At present, periodic systems can be calculated only with DFTB. In PBC, the periodic cell should be neutral, although the summation of the electrostatics adopted in DFTB/PBC appears to avoid divergence for charged cells via the use of tin foil conditions.

FMO/PBC cannot use any rotational symmetry (that is, the periodic cell in FMO should be defined for P1 space group), which may require adding symmetry-related copies of the elementary cell into a cell that has only periodic symmetry.

For fragments not connected by covalent bonds, there is nothing special in terms of fragmentation. Pure liquids, their mixtures, and solutions can be calculated by constructing an orthogonal periodic box of a suitable size. The size of a box for a solution determines the concentration of the solute. Counterions in solutions may be used with care (Section 4.11.5). Non-orthogonal cells may be used for molecular crystals.

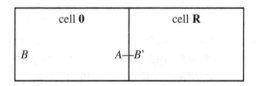

Figure 25.   Example of a bond between atoms A and B that should be declared in $FMOBND because an image of B forms a bond with A, $A–B'$.

Fragmenting periodic systems in solid-state materials with detached bonds between fragments is tricky to accomplish. A pair of atoms A and B can have a bond to be detached in $FMOBND, for $A \in \mathbf{0}$ (original cell) and $B \in \mathbf{R}$ (a replicated cell), as shown in Figure 25.

CHAPTER 5

Advanced Techniques

5.1 How to Compute Polarization Energies

Three different approaches can be pursued to obtain polarization energies: PIEDA, AP method, and interaction energy analysis (IEA) for solute/solvent polarizations. There peculiarities are shown in Table 71. The most commonly used method is PIEDA. A detailed description of each method follows.

Table 71. Peculiarities of methods for computing polarization (PL) energies.

Variety	System	Solvent[a]	Caps[b]	Reference
PIEDA	general	FMO	yes	isolated
AP	general	PCM/SMD	no	isolated
IEA	solution	EFP	no	vacuum

[a]Solvent may be omitted in PIEDA and AP.
[b]Minimal caps for computing the isolated state of fragments with detached covalent bonds.

5.1.1 Stepwise Polarization in PIEDA

The physical concept of polarization is described in Section 3.4. To compute polarization energies of fragments, it is necessary to do several calculations (Table 72). These steps should be done in the order shown, by passing the results from one step to the next. In PL0 and PL calculations it is necessary to use PIEDA (but not in the isolated state). Empirical dispersion models such as D3(BJ) are not supported in a polarization calculation (and they make no effect anyway because they are not polarizable). In addition, continuum solvent models (PCM and SMD) may not be used. Explicit solvent may be used as FMO fragments.

The isolated state of a fragment is obtained without embedding from other fragments. This state is straightforward to compute if there are no covalent bonds to other fragments (e.g., for a water molecule). For a fragment connected by covalent bonds to another fragment, the

Table 72.   Step-by-step guide to computing polarization in PIEDA.

Step	Input data	Results
0 state	(structure)	Energies E_I^0 (DAT file) and densities $\mathbf{D}_I^0$ (F30 file) of isolated fragments.
PL0 state	E_I^0 as EFMO0,[a] $\mathbf{D}_I^0$ as file F40.	Polarization energies ΔE_I^{PL0} for the fixed embedding, and PL0 couplings (DAT file).
PL state	E_I^0 as EFMO0.[a] ΔE_I^{PL0} as EPL0DS.[a] ΔE^{A0} (A = ES, etc.) as EINT0.[a]	Polarization couplings $\Delta E_I^{PL\text{-}PL}$ and PL couplings of polarization to other components in PIEDA (output file).

[a]A keyword in $FMOPRP.

isolated state is defined as the electronic state of the fragment with minimal caps.

Consider a residue fragment in a protein, for example, alanine, which in FMO looks like $-CO-NH-HC_\alpha Me-$. Minimal methyl (Me) caps are added, resulting in **Me**$-CO-NH-HC_\alpha Me-$**Me**, divided into three fragments, two methyls and alanine. Then FMO calculations are conducted, and the energy and density of the alanine fragment are taken to represent its isolated state. The algorithm for an automatic addition of caps is quite basic and may not properly work for complicated cases. It is not possible to modify the automatically determined caps.

In these isolated calculations, F40 is overwritten for each fragment, and the isolated state of all fragments is stored as F30. It is necessary to modify rungms to prevent it from deleting F30. The

Table 73. Input header for an isolated (0) state calculation.

Input	Comments
⌐ $CONTRL RUNTYP = FMO0 $END	A special RUNTYP for FMO0.
⌐ $FMOPRP MODORB = 3	Complete format of F40/F30.
MAXAOC = M	For writing out F30.
COROFF = 0 $END	In DFT, turn off switching to RHF. [a]

[a]Needed because the embedding loop is not used (in FMO0, SWOFF in $DFT may be used instead of COROFF).

isolated state calculation is accomplished with RUNTYP = FMO0, and a sample input header is shown in Table 73.

In an isolated state calculation, MAXAOC should be set. Its value M is equal to max$\{M_I\}$, where M_I is the number of AOs in fragment I. Practically, the easiest way to get this number may be to run an FMO calculation (other than FMO0) and find the following line "Max AOs per n-mer:". The first number printed after it is M.

For PBC, the recommended isolated state is for a single fragment with PBC (such a calculation includes the self-polarization by the images of the "isolated" fragment). To do this calculation, no special options are needed in FMO0, and only a PBC cell should be defined. Alternatively, one could use the isolated state of a molecule without PBC, and evaluate the self-polarization by comparing the isolated states with PBC and without. For this path, an extra step in the PIEDA scheme is needed (FMO0 without PBC).

Suppose that the input file for the 0 state is called $job.0.inp and for the PL0 state $job.pl0.inp. After the isolated state is computed, $job.0.F30.000 should be renamed as $job.pl0.F40.000 and the file should be copied to the run-time directory of the master node prior to running the PL0 job as a restart.

A PL0 calculation is computed in the fixed embedding due to the densities of isolated fragments. The input file header for a PL0 calculation is shown in Table 74. As a result, one-body polarization energies of fragments are obtained. IPIEDA for both PL0 and PL states is set to 2 when polarization is computed; for PL0, MAXIT = 1 in $FMOPRP is also set.

Table 74.   **Input header for a PIEDA/PL0 calculation.**

Input	Comments
⌴$FMOPRP MODORB = 3	Complete format of F40.
COROFF = 0	In DFT, turn off switching to RHF.[a]
MAXIT = 1 IPIEDA = 2	PL0 state.
IREST = 2 MODPAR = 77	F40 is read on rank 0 and broadcast.
EFMO0(1) = −76.2157137341…	Isolated fragment energies (from DAT file of 0 state).
ROBDA(1) = 1.52459598,…	Optional, if BDA corrections are to be used.[b]
EOBDA(1) = −14.5205976043,…	Optional, if BDA corrections are to be used.
⌴$END	

[a]Needed because only 1 iteration of the monomer loop is performed (fixed embedding).
[b]Note that NOBDA should not be set.

Table 75.   **Input header for a PIEDA/PL calculation with polarization energies.**

Input	Comments
⌴$FMOPRP MODORB = 3	Complete format of F40.
COROFF = 1e-3	Default value (0 may also be used).
IPIEDA = 2	PL state using results of 0 and PL0 states (note no MAXIT).
EFMO0(1) = −76.2157137341…	Isolated fragment energies (from the DAT of 0 state).
EPL0DS(1) = 0.0044847113…	Polarization components from the DAT of PL0 state.
EINT0(1) = −33.6738801458…	Sums of PIEDA/PL0 components from the DAT of PL0 state.
	(Optionally, ROBDA and EOBDA for BDA corrections).
⌴$END	

The last step is to copy the results of 0 and PL0 states to the input file of the final PL calculation. The PL state does not require a restart. A sample header is shown in Table 75. As a result, both destabilization and stabilization polarization couplings are obtained.

In addition to polarization, PL0 and PL state calculations deliver various couplings of PIEDA terms (Section 3.5.5) to polarization, such as polarization-exchange coupling. In PL0 and PL calculations, the energies of BDA corrections (Section 4.11.3) can be provided by defining R0BDA and E0BDA.

5.1.2 *Polarization in AP*

The AP method is described at length in Section 5.4. There are two ways to analyze AP results: either as a basis set correction or as polarization. Namely, the energy in AP is computed for two basis sets BS1 (medium) and BS2 (large) with embedding (V) or without it (0).

$$
\begin{aligned}
E^{\text{FMO/AP}} &= E^{\text{FMO}}(\text{BS1},V) - E^{\text{FMO}}(\text{BS1},0) + E^{\text{FMO}}(\text{BS2},0) \\
&= E^{\text{FMO}}(\text{BS1},V) + \Delta E^{\text{BS}} \qquad\qquad (86) \\
&= E^{\text{FMO}}(\text{BS2},0) + \Delta E^{\text{PL}}
\end{aligned}
$$

One interpretation of FMO/AP is that it is a regular FMO calculation with embedding and a medium basis set with the energy of $E^{\text{FMO}}(BS1,V)$ plus a basis set correction, defined by

$$
\Delta E^{\text{BS}} = E^{\text{FMO}}(\text{BS2},0) - E^{\text{EMO}}(\text{BS1},0) \qquad\qquad (87)
$$

Another view of FMO/AP is that it is a large basis set calculation without embedding with the energy of $E^{\text{FMO}}(\text{BS2},0)$ plus a polarization correction, defined by

$$
\Delta E^{\text{PL}} = E^{\text{FMO}}(\text{BS1},V) - E^{\text{FMO}}(\text{BS1},0) \qquad\qquad (88)
$$

By conducting an AP calculation for FMOn, polarization energies $\Delta E^{\text{PL}m}$ are obtained for $m \leq n$. $\Delta E^{\text{PL}1}$ describes the polarization effects of monomers, that is, it corresponds to the monomer destabilization in PIEDA (but note that for covalent boundaries the definition of the isolated state differs: in PIEDA there are minimal caps, whereas in AP there are no caps). Higher order terms $\Delta E^{\text{PL}m}$ for $m>1$ include the stabilization part of the polarization.

Decomposing each FMO energy in Eq. (86) in the many-body series, one obtains pair interaction energies for AP. They can also be interpreted in two ways, analogously to the total energies,

$$
\begin{aligned}
\Delta E_{IJ}^{AP} &= \Delta E_{IJ}^{BS1,V} - \Delta E_{IJ}^{BS1,0} + \Delta E_{IJ}^{BS2,0} \\
&= \Delta E_{IJ}^{BS1,V} + \Delta E_{IJ}^{BS} \\
&= \Delta E_{IJ}^{BS2,0} + \Delta E_{IJ}^{PL}
\end{aligned}
\tag{89}
$$

ΔE_{IJ}^{BS} is the difference of two PIEs for (BS2,0) and (BS1,0), each of which can be decomposed in PIEDA, producing a BS correction to each of the PIEDA components. For instance, taking the MP2 decomposition in Eq. (28),

$$
\begin{aligned}
\Delta E_{IJ}^{BS} &= \Delta E_{IJ}^{BS2,0} - \Delta E_{IJ}^{BS1,0} \\
&= \Delta E_{IJ}^{ES\cdot BS} + \Delta E_{IJ}^{EX\cdot BS} + \Delta E_{IJ}^{CTMIX\cdot BS} + \Delta E_{IJ}^{RCDI\cdot BS} + \Delta E_{IJ}^{solv\cdot BS}
\end{aligned}
\tag{90}
$$

where the basis set correction to each component A (A = ES, etc.) is

$$
\Delta E_{IJ}^{A\cdot BS} = \Delta E_{IJ}^{A,BS2,0} - \Delta E_{IJ}^{A,BS1,0}
$$

Likewise, one can define the polarization coupling to each PIEDA component as

$$
\begin{aligned}
\Delta E_{IJ}^{PL} &= \Delta E_{IJ}^{BS1,V} - \Delta E_{IJ}^{BS1,0} \\
&= \Delta E_{IJ}^{ES\cdot PL} + \Delta E_{IJ}^{EX\cdot PL} + \Delta E_{IJ}^{CTMIX\cdot PL} + \Delta E_{IJ}^{RCDI\cdot PL} + \Delta E_{IJ}^{solv\cdot PL}
\end{aligned}
\tag{91}
$$

where the polarization effect on each component A (A = ES, etc.) is

$$
\Delta E_{IJ}^{A\cdot PL} = \Delta E_{IJ}^{A,BS1,V} - \Delta E_{IJ}^{A,BS1,0}
$$

5.1.3 *Polarization in Solution (IEA)*

In order to analyze solvent effects for QM/EFP, IEA can be performed, treating solvent molecules with EFP. IEA can also be applied to unfragmented calculations set up as FMO with 1 fragment.

To compute polarization with IEA, one has to define a reference state. A useful reference state for IEA is a vacuum calculation. This corresponds to the isolated state in PIEDA. The density of this state can be read and solute-solvent interactions can be obtained, similarly to the PL0 state. Finally, polarization couplings are obtained by doing a normal polarized (PL) state. Thus, a polarization calculation in IEA consists of essentially the same 3 steps as for PIEDA with polarization.

The solvation energy ΔE^{solv} is obtained by subtracting the pure solute (FMO in vacuum) and pure solvent (EFP) energies from the energy of the solvated solute (FMO/EFP),

$$\Delta E^{solv} = \Delta E^{FMO/EFP} - \Delta E^{FMO} - \Delta E^{EFP}$$
$$= \sum_{I=1}^{N} \left(\Delta E_I^{es,0} + \Delta E_I^{rem,0} + \Delta E_I^{FMO,pol} + \Delta E_I^{EFP,pol} \right) + \Delta E^{FMO-EFP,pol} \quad (92)$$

where $\Delta E_I^{es,0}$ is the electrostatic solute-solvent interaction, computed for the reference state of the solute (for simplicity, the structure deformation between vacuum and solvent is neglected in Eq. (92)), and $\Delta E_I^{rem,0}$ is a similar non-electrostatic interaction term, called the remainder interaction. $\Delta E_I^{FMO,pol}$ is the polarization of FMO fragment I by all EFP fragments. $\Delta E_I^{EFP,pol}$ is the contribution of FMO fragment I to the polarization of all EFP fragments. $\Delta E^{FMO-EFP,pol}$ is the many-body solute-solvent polarization coupling (corresponding to $\Delta E^{PL\text{-}PL}$ in PIEDA).

It is possible to decompose total EFP contributions into individual EFP fragment values. For example, $\Delta E_I^{es,0}$ can be decomposed into a sum of $\Delta E_{IJ}^{es,0}$ (for FMO fragment I and EFP fragment J), and these values correspond to ΔE_{IJ}^{ES0} in PIEDA with water treated as QM fragments (Section 5.1.1). Likewise, $\Delta E_I^{rem,0}$ can be decomposed into a sum of $\Delta E_{IJ}^{rem,0}$, which correspond to $\Delta E_{IJ}^{EX0} + \Delta E_{IJ}^{CT\ mix0}$ in PIEDA.

Using the multipole expansion of the electrostatic interaction in EFP, the contributions of monopoles (charges), dipoles, quadrupoles, and octupoles can be obtained. $\Delta E_I^{FMO,pol}$, similar to Eq. (92), can be decomposed into destabilization and stabilization components.

Table 76.   Options for IEA in $FMOEFP.

Keyword	Meaning
IEACAL = 1	Perform IEA.
ITRLVL = 1	In IEA, use free state (usually, with a restart IREST = 2 MAXIT = 1).
ITRLVL = 2	In IEA, use polarized state.
IEABDY = m	In IEA, treat up to m-body terms ($m \leq n$ for FMOn).
NPRIEA = X	A packed option for print-out in IEA.[a]
IPFMO	Array listing indices of FMO fragments for which IEA is done.[b]
IPEFP	Array listing indices of EFP fragments for which IEA is done.[b]

[a]See docs-input.txt for available choices.
[b]The default is to use all fragments, so these options may be omitted.

Polarization couplings of PIEDA components are obtained in IEA, by subtracting the values of PIE in vacuum and in solution.

The options for IEA are shown in Table 76. An IEA calculation can be accomplished by doing three steps: (1) vacuum FMO of the solute (saving F40), (2) FMO/EFP with ITRLVL = 1 to get $\Delta E_I^{es,0}$ and $\Delta E_I^{rem,0}$ (reading F40 in a restart), and (3) a regular FMO/EFP run with ITRLVL = 2 to get the rest of the components.

There are two modifications of IEA, different in the scaling factor of the dimer polarization contributions: 1 for IEACAL = −1 (the original version), and 2 for IEACAL = 1 (the newer improved version).

5.2 Embedding Types

Fragment calculations are usually performed in an embedding ESP, or an embedding for short. The potential has two important roles: (a) to describe the polarization and (b) to saturate covalent bonds at the fragment boundaries.

In an unfragmented calculation, there are no hydrogen caps on bonds. This works because electrons feel the appropriate potential, so there is no need to add any hydrogen caps because bonds are "saturated" by the potential.

In FMO there are no hydrogen caps at fragment boundaries and a proper distribution of the electron density is accomplished by the

Table 77.   Embeddings in FMO.

Embedding	Input options	Usage
None	RUNTYP = FMO0	Compute isolated fragments.[a]
Exact ESP	RESPPC = 0	Rarely used (e.g., for MP2 with SCZV).
Hybrid ESP[b]	RESPPC = R	Most commonly used ESP.
Point charge ESP	RESPPC = −1	(a) Standard embedding in DFTB,[c] (b) Extended basis sets (Section 5.2), (c) Polarizable+fixed embedding.
Damping for point charges	SCREEN(1) = 1,1	Extended basis sets (Section 5.2).

[a]Strictly speaking, for connected fragments in FMO0, the embedding of minimal caps is used.
[b]Exact ESP for near fragments and point charge ESP for fragments separated by R or more (R is usually 2.0). RESPAP = A can be added for an intermediate approximation, where A is usually 1.0 (see Figure 8).
[c]The value of RESPPC is ignored in DFTB, but it should be non-zero.

embedding potential plus an additional potential for fragment boundaries. This latter potential is just a way to divide redundant AOs across a boundary, so it is not really an embedding per se.

The form of the embedding is very important, and strongly affects the results (Table 77). An embedding can be used in a fixed or self-consistent way. In most cases, the latter is used, and the former is employed mainly in the calculation of polarization energies (Section 5.1).

The exact ESP, computed from electron density, does not necessarily give a better accuracy than other ESPs, because the accuracy in FMO is not straightforwardly determined by how faithfully its ESP reproduces the potential in full calculations. The recommended embedding is the hybrid ESP (point charges are used for far fragments and the exact embedding is used for close fragments) except that in DFTB only a point charge potential can be used.

The embedding in FMO is computed from the electron densities obtained in SCF. For example, in MP2, the embedding is computed with the electron density of HF, just as in unfragmented calculations. Even though there may be an MP2 electron density, it is never used to get the embedding potential. Likewise, in TDDFT, the ground state

DFT electron density is used to get the embedding. In MCSCF, the electron density from MCSCF is used to compute the embedding due to the MCSCF fragment (possibly for an excited state).

The atomic charges, which are used for the point charge embedding, can be of various types: (a) Mulliken, (b) Stone, and (c) so-called ESP charges. The analytic gradients are available only for the most commonly used Mulliken charges. Stone charges are from the Stone multipole analysis, and ESP charges are fitted to reproduce the ESP due to the electron density on a set of grid points. The accuracy of FMO is similar for various charge types, and Mulliken charges are typically used.

To correct for charge penetration of electron clouds, a damping can be used in ESP of point charges by setting SCREEN (1) = 1,1, which has some beneficial effect for metal ions. There are no analytic gradients for damped charges. DFTB has a different damping via the use of γ and Γ functions (with analytic gradients).

The hybrid embedding suffers from an accuracy loss in the application of the RESPPC approximation to the embedding potential acting on monomers, dimers, and trimers. For example, in a linear chain of fragments $A...B...C$, with AB and BC separations slightly less than RESPPC, the ESP due to C is computed exactly for dimer AB (C is close to AB), but approximately for monomer A (C is far from A).

For FMO2 this inconsistency problem is very minor (owing to the formulation of the ESP term in Eq. (6)), but for FMO3 the accuracy loss is more noticeable. A balanced blockwise scheme can be used for ESP, in which the application of approximations in dimers and trimers is done for each monomer block in the ESP matrix separately. It is turned on with MODESP (usually MODESP = 1). There is no analytic gradient for blockwise ESP. The FMO3 accuracy can be improved with this scheme.

A fixed embedding can be used for some fragments if RESPPC = −1 is set, to model electron excitations in a quasi-crystal field for a large finite cluster representing a molecular crystal. This technique can be used for other purposes, whenever the user wants to add a fixed (non-polarizable) field of point atomic charges. These fixed charges can be defined for some fragments in FMO (the whole fragment can be defined as polarizable or frozen), that is, it is not a general way to define arbitrary point charges, but rather, a way to use a fixed point charge embedding due to some fragments.

SCF can be skipped for selected fragments, and the embedding due to them computed with user-specified point charges. For a quasi-crystal embedding, one can compute atomic charges in DFT/PBC in an external program and use them as fixed charges for a large cluster of atoms representing the crystal, except a central (polarizable) QM part, which is computed with FMO. Dimers and trimers including fixed embedding fragments are not computed. This is a rudimentary way of doing electronic embedding in QM/MM for FMO. There is no analytic gradient for this scheme.

The use of fixed embedding is accomplished with SCFFRG by assigning NONE to those fragments for which fixed charges are to be used. There are two ways to define the values of fixed charges.

In the general approach, the charges are set up as ATCHRG. The order of charges, however, is not the same as for atoms in $FMOXYZ. In ATCHRG, the charges are specified for atoms in fragment 1, then in fragment 2, etc., in the exact same order as the atoms appear in each fragment. Atomic charges for BDAs, shared between two fragments, should be defined in each of the two fragments, split in a way consistent with the division of electrons at the boundary (Section 4.11).

The second approach is for a fixed embedding due to identical molecules (such as solvent molecules in a liquid or monomers in a molecular crystal). All fragments with fixed charges should have the same order of atoms. In this approach, ATCHRG defines the charges of a single fragment, and NATCHA defines the number of atoms in it. These charges are used for those fragments I for which 64 is added to NOPFRG(I). All fixed charge fragments should have NATCHA atoms.

An example of using this feature is $FMO RESPPC = −1 SCFFRG(4) = NONE,NONE NOPFRG(4) = 64,64 NATCHA = 3 ATCHRG(1) = −0.8,0.4,0.4 $END. Here, two water fragments, numbers 4 and 5, are treated as fixed charges. The values of charges are in ATCHRG. In this example, both fixed fragments use the same charges. It is necessary to use the point charge embedding (RESPPC = −1).

5.3 Diffuse Basis Sets

The accuracy of FMO and SCF convergence degrade as the basis set becomes more diffuse. Pople triple-ζ with polarization may still be used with the default embedding (preferably with FMO3), but Dunning cc-pVTZ and any basis sets with diffuse functions (such as 6-31+G*) require a special treatment. Users are strongly discouraged from simply specifying such a basis set without taking one of the measures described below.

For diffuse basis sets, it is recommended to set QMTTOL in $CONTRL to remove linear dependencies; sensible choices are 1e-5 or 1e-6. Poor convergence, if caused by a linear dependency alone, can be cured by increasing the value of QMTTOL, which, however, should not be set to too large a value either (the user should be reluctant to increase it beyond 1e-5). In addition, ITOL = 24 and ICUT = 12 should be set in $CONTRL to increase the accuracy for computing the integrals; using the default values may result in SCF not fully converging. DIRTHR in $SCF is sometimes helpful in these cases.

If diffuse functions are important only for a small local part of a molecular system, all such atoms can be put into one fragment, and a multiple basis set calculation can be attempted. This approach may face convergence problems, if diffuse functions are used for more than a few atoms.

If there are no covalent boundaries between fragments, then the point charge embedding of RESPPC = −1 can be used. This embedding works quite well with diffuse functions; its drawbacks are that it is inaccurate for covalent boundaries and it has no exact gradients. If one is to compute a cluster of molecules with MP2 or CCSD(T) and an extensive basis set, this point charge embedding may work well. Geometry optimizations or MD should not be done with this approach, due to the gradient inaccuracy.

A more general method for diffuse basis sets is AP (Section 5.3). It can be used with covalent boundaries and it has the exact gradient. AP is usually used with the default embedding.

Diffuse functions in a basis set are often used to reduce BSSE. There are two alternative ways to deal with BSSE. The main part of

BSSE comes from core electrons, which are of little importance to chemistry directly. Thus, the alternative treatment is to remove core electrons with a core potential (see Section 4.1.5). Note that in DFTB, core electrons are always removed. Alternatively, HF-3c can be used (Section 4.1.2) with two special parametrized corrections for BSSE.

5.4 Dual Basis Sets (AP)

AP is a dual basis set approach, which is used for large basis sets. The smaller basis set is used to compute the polarization; the larger basis set is used to add a BSSE correction. Polarization and BSSE corrections are computed separately and added up, not unlike other additive schemes in quantum chemistry.

In AP, three calculations are done: (1) medium basis set without embedding, (2) medium basis set with embedding, and (3) large basis set without embedding, as a way of avoiding the use of the large basis set with embedding, which is difficult in FMO. There are two conceptual ways of looking at AP: (a) a BSSE correction to the regular FMO calculation, i.e., $(2)+[(3)-(1)]$, or (b) a polarization correction added to a calculation with the large basis set and no embedding, as $(3)+[(2)-(1)]$, which yield the same total result.

There is an analytic gradient for AP but no analytic Hessian. AP can be combined with PIEDA and PCM.

A typical usage of AP is to employ a double-ζ basis set such as 6-31G** (or cc-pVDZ) as the medium basis set and add diffuse functions in the large basis set 6-31++G** (or aug-cc-pVDZ). Another usage might be to combine cc-pVDZ (medium) with cc-pVTZ (large). There is no formal limitation of mixing basis sets, so in principle one could use both Pople (medium) and Dunning (large) basis sets.

Convergence problems may arise in AP, typical for large basis set (see Section 5.3), which, however, are often surmountable with the removal of linear dependencies (QMTTOL = 1e-5) and tricks to deal with bad convergence (Section 4.10).

To set up an AP calculation, NDUALB = 1 is added to $FMO and two basis sets have to be defined in the dual basis set fashion (as multiple basis sets in 1 layer), Section 2.6.1 (note that $FMOXYZ should not include basis set extensions in the names of atoms). For each basis set, hybrid orbitals have to be specified in $FMOHYB if there are covalent boundaries between fragments. For each entry in $FMOBND, two HMO set names should be listed, the first for the smaller and the other for the larger basis set. Normally, AP is used with the default (hybrid) ESP.

5.5 Restarting Jobs

5.5.1 *Geometry Restarts*

Simple restarts can be done for RUNTYP = OPTIMIZE, OPTFMO, SADPOINT, and MEX. The easiest is to take the last geometry from a previous output file and put it into $FMOXYZ. Note that this will reset the Hessian, so that a geometry optimization will not restart exactly. A smooth restart is possible for OPTIMIZE (by pasting the Hessian from DAT file) and OPTFMO (by pasting the $OPTRST group from DAT file). If IREST = 0 in $OPTFMO is set, no restart data are punched.

Semi-numerical Hessian restarts are possible for RUNTYP = HESSIAN (but not FMOHESS). These restarts can be done by copying relevant data (gradients and dipole moments) from the DAT file of a preceding run into the input file.

5.5.2 *MD Restarts*

In MD, a restart information is written once in KEVERY steps, so if a trajectory ran for 1245 points with KEVERY = 100, it can be restarted from step 1200. For a restart, replace the $MD group by the data in the restart file generated by a preceding MD (a restart $MD group will have proper options including an initial velocity of each atom).

Some manual editing of the input file is necessary. Atomic coordinates in $FMOXYZ have to be replaced with coordinates found in the restart file. NSTEP (the number of steps in MD) should be adjusted. Some advanced MD options may not be properly written to the restart file, for example, umbrella sampling options are written with 4 significant figures in the restart file, and if the original value had more, they have to be restored. One should carefully look over all options written in the $MD group in the trajectory file to make sure that they are correct.

For PBC, if the default atom wrapping is disabled, atoms may spill out of the original cell in MD. In this case, for a restart, NSPILL should be set. For example, NSPILL = 2 is used when some atoms wandered two cells away from the original cell in any of the three directions.

5.5.3 *Fragment-Specific Text Restarts*

It is possible to read initial MOs for selected monomers, dimers, or trimers. For example, to provide initial orbitals for fragments 2 and 5, and dimer 5,4, the user should put their MOs as separate VECi$ groups: for example, MOs for fragment 2 as $VEC1, for fragment 5 as $VEC2, and for dimer 5,4 as $VEC3. These MOs can be generated by a preceding FMO run, in which the user took care to punch MOs by using NPUNCH = 2 in $SCF.

As an example of a restart usage, $FMOPRP IJVEC(1) = 2,0,0,1,35, 5,0,0,1,46, 5,4,0,1,70 $END is taken. Entries in IJVEC come in quintuplets, one per VECi$ group, in the order of increasing i. Each quintuplet has the format: I,J,K,L,M, where I,J,K describe an n-mer: for monomers, $I,0,0$; for dimers $I,J,0$ and for trimers I,J,K (I,J,K must be listed in the decreasing order; use 5,4,0, and not 4,5,0). L is the layer for which the orbitals are read. M is the number of MOs to read. It is possible to read an incomplete set but it should include at least all occupied orbitals. In the above example, 35 MOs are read for fragment 2, 46 for fragment 5, and 70 for dimer 5,4 (in layer $L = 1$). For UHF (UDFT), VECi$ should have M α MOs, followed by M β MOs.

5.5.4 *All-Fragment Binary Restarts*

A binary restart is done for PIEDA/PL0 and it is rarely useful in other runs. In this type of restart the electronic state of all fragments is read and reused. To do this, restart file $job.F40.000 from a previous run is reused (this file is typically erased by rungms). Some runs do not generate this file (e.g., for MODIO = 3072 in $SYSTEM or MODPAR = 1024 in $FMOPRP). The format of F40 depends on MODORB and SCFTYP, and a restart is possible when these are the same in the two runs.

In some tasks, there is a file F30 that has the same structure as F40. For example, in RUNTYP = FMO0, F40 is recreated and destroyed for each fragment, whereas F30 is the global file for all fragments. F30 from FMO0 should be saved, renamed as F40, and used in a restart for PL0. For an orbital conversion in MCSCF, likewise, F30 for MCSCF is created from F40 of RHF, and F30 should be then renamed to F40 for MCSCF.

To do an F40 restart, $job.F40.000 from a previous run should be copied to the run-time scratch directory SCR of the master node prior to executing rungms. $FMOPRP IREST = 2 MODPAR = 77 should be added to the input file for restart. IREST = 2 means that the restart is for monomers (step 2 in Table 1). MODPAR = 77 (64+8+4+1) is the default option 13 plus 64, where the meaning of 64 is to broadcast the contents of F40 by rank 0 to all other ranks.

Some advanced ways of running FMO cannot be properly restarted, such as AP, and some tasks, such as FD, have additional complications. Restarting from the dimer or trimer step is possible, but complicated.

5.6 Using FMO for Non-FMO Tasks

FMO can be used for generating useful data for non-FMO calculations, if SCF in them diverges, as may happen in a protein. To generate an initial electron density for a non-FMO run using FMO, MODPRP = 1 is set in $FMOPRP, and file F10 from FMO calculations should be saved (rungms deletes it). For this initial guess in a non-FMO calculation, GUESS = MOREAD in $GUESS is used and the saved file F10 should be copied to the run-time SCR directory of the master node before executing rungms. In GDDI, F10 gets a rank extension, which is not added in DDI.

This F10 has only density but no MOs, so it can be used for RHF or RMP2, but not for DFT or MCSCF (to use it in DFT, SWOFF = 1e-3 in $DFT may be tried).

If a Hessian is needed to locate a transition state for an enzyme in unfragmented calculations, and the Hessian is too expensive, FMO can be used to calculate an initial Hessian, handy for a geometry optimization or a transition state search in unfragmented calculations.

Some tasks are not possible without FMO, so if one wishes to do them for an unfragmented calculation, the only way is to set up FMO with 1 fragment. These features include partition analysis, IEA, and geometry optimizations of lattice vectors using RUNTYP = OPTFMO.

For such non-FMO runs masqueraded as a 1-fragment FMO, an input file should include $FMO NFRAG = 1 NBODY = 1 NACUT = M MODGRD = 0 $END, where M is the total number of atoms. For DFT, COROFF = 0 should be set in $FMOPRP. INDAT or other fragmentation arrays are not needed; however, a charge and a multiplicity have to be specified in $FMO as ICHARG and MULT, respectively. SCF and DFT types and other QM details can be given in general groups, or in $FMO. In PCM, $PCM IFMO=0 $END should be used.

5.7 Temperature, Entropy, and Free Energy

For a single minimum, temperature-dependent vibrational entropy, enthalpy, and free energy can be calculated using thermodynamics, via a Hessian calculation, either analytic or semi-numerical. In the latter case, one can use PAVE (Section 3.5.7) for a convenient fragmentation and analysis. Such calculations can treat vibrational contributions for one minimum, but not conformational entropy. Up to 10 temperature values may be specified as TEMP in $FORCE.

The alternative is to use FMO/MD simulations. Currently, only umbrella sampling is readily available. By doing MD simulations, and processing the results, one can obtain free energy, as described in Section 3.10.2 for chemical reactions. A similar method may be used for binding processes; however, defining an umbrella-sampling coordinate may not be feasible, so a different way to constrain the binding coordinate should be devised.

Some add-on models used in FMO have temperature-dependent parameters, most notably ICAV in PCM. The temperature can be specified as TABS in $PCM, affecting the solvation energies.

5.8 Acceleration Tricks

Some acceleration techniques are so general that they are executed by default. Some, however, are not, mainly for two reasons: (1) by gaining speed one loses something; usually, some property computation is skipped and it must be an explicit wish of the user to do so, and (2) acceleration options are implemented with various restrictions and may disrupt calculations if misused.

Whether or not a particular option makes a substantial effect on performance depends on the calculation; sometimes the effect is small and sometimes huge. These options are meant for expert users, who are capable of a critical assessment of results and spotting suspicious behavior.

Some ways to reduce the output are also described, which may be very handy during MD and geometry optimizations. If there is any problem, like a convergence issue, it is usually necessary to turn off output reduction options to gain as much information as possible.

There are two "superoptions" provided, set in the packed MODIO in $SYSTEM, 1024 and 2048, which are usually used together as MODIO = 3072. The first (1024) turns on some acceleration options, and the other (2048) reduces the output. This combined superoption of MODIO = 3072 is well tested and recommended for geometry optimizations, MD simulations, and semi-numerical Hessians. It is in general not advised to use these superoptions for single-point runs.

5.8.1 *Acceleration of I/O*

A detailed output file in FMO can become gigantic, and reducing its size can improve the efficiency. NPRINT in $FMOPRP is a rare example of a packed option, in which 2 bits are used together rather than the usual 1 bit per suboption. The initial two bits of NPRINT define the main print-out level for FMO, which can be 0, 1, 2, or 3, with a decreasing amount of output. Single-point calculations often use NPRINT = 1; geometry optimizations and MD use NPRINT = 3. Adding 128 to NPRINT prevents printing pair contributions for separated dimers (which can be very numerous). It does not affect the results, as these contributions are calculated and added to the total properties but not printed as a table at the end.

A summary of I/O related options is given in Table 78. Several options require a guess of the necessary memory size (an upper estimate), such as MEM10.

Simple as it is, not setting DIRSCF = .T. in $SCF (the default is .F.) can have a disastrous effect on performance of deeply shocking proportions if the run-time directory SCR is on a network disk (incidentally the default!).

Table 78.   I/O related acceleration options.

Group	Option	Meaning
$CONTRL	NPRINT = −5	Stop printing MOs of fragments in the output.
$SCF	DIRSCF = .T.	Direct SCF (integrals are not stored on disk).
$SCF	NPUNCH = 0	Stop punching MOs of fragments ($VEC).
$SYSTEM	MEM10 = L[a]	Reserve L words for storing dictionary file F10 in memory.
$SYSTEM	MEM22 = L[a]	Reserve L words for storing file F22 (used in DFT) in memory.
$SYSTEM	MODIO = 4096	Redirect output from all groups except group 0 to /dev/null.[b]
$FMOPRP	NPRINT = 131	Puny output level and no output of ES dimer energies.
$FMOPRP	MODPAR = 1033	Store fragment data in memory rather than on disk (F40).
$OPTFMO	MAXNAT = K[c]	DFTB restart file is stored in memory.
$GDDI	PAROUT = .F.	Output and DAT files are opened only by group masters.

[a]Note that the units are words, not megawords. 10000000 is a typical guess of L.
[b]It may scrap important error messages.
[c]K should be set to the maximum number of atoms per fragment.

In FMO calculations, fragment data (fragment densities or MOs, as controlled by MODORB in $FMOPRP) are stored on a separate file F40 by every compute process. It is possible to use shared memory for this information by adding 1024 to MODPAR in $FMOPRP, typically, as MODPAR = 1033. Such storage in memory cannot be used with some FMO tasks, for example, polarization. Data for dimers and trimers are never stored on F40.

There is a radical thing one can do in UNIX about I/O, which is to use the electronic disk. There is a directory (/dev/shm) that is actually a part of memory, not hard disk. It is possible to use it for storing run-time files (by defining it as SCR in rungms) provided that there is enough physical memory for both GAMESS executable and these files. A misplanned usage of more memory than is available can degrade performance. Note that after a reboot or shutdown all information on the electronic disk will be lost. If a node is shared with other users, they may not like that their memory is clandestinely used in this way.

5.8.2 *Acceleration of Data Processing*

The options to accelerate data processing are summarized in Table 79. KDIAG in $SYSTEM determines the matrix diagonalization algorithm, which is important for DFTB (fragment Fock matrix diagonalization), geometry optimization with RUNTYP = OPTIMIZE (Hessian diagonalization, note also the option KDIAGH in $FORCE), or LCMO (total Fock matrix diagonalization), where the matrix diagonalization can be a major bottleneck. The default method KDIAG = 0 has some parallelization, but it is not very efficient. A good alternative is to use KDIAG = 5 if GAMESS is compiled with a supported LAPACK library (that is, MKL).

NINTIC = K in $INTGRL can be used with conventional SCF (DIRSCF = .F. in $SCF), in which integrals are precomputed before SCF begins. Ideally, K should be large enough to fit all integrals in memory, in which case there may be some acceleration gained vs. direct SCF. If the allotted memory is not enough, some integrals will be stored on disk, slowing down the calculation. NAODIR in

Table 79. Acceleration options for a more efficient data processing.

Group	Option	Meaning
$SYSTEM	KDIAG = 5[a]	Fast matrix diagonalization.
$PCM	MODPAR = 64[b]	Prestore PCM data for solving Eq. (22).
$PCM	MODPAR = 256	Use ASCs from previous geometry in PCM.
$INTGRL	NINTIC = K[c]	Use replicated memory to store two-electron integrals.
$DFTB	MODGAM = 13	Accelerate electrostatics for γ and Γ-functions in DFTB.
$DFTB	MODESD = 2	Accelerate separated dimers in DFTB.
$FMOPRP	NGUESS = 8[b]	Reuse dimer data for geometry optimizations and MD.
$FMOPRP	PRTDST(1) = 0,0, −1,0	Turn off geometry and fragmentation checks.
$FMOPRP	MODCHA = 2	Skip three-body CT calculation in FMO3.
$FMOPRP	MODPRP = 4096	Accelerate FMO3 (requires turning off suboption 1 in MODPAR).
$DFT	SG1 = .T.	Use fast grid in DFT.
$TDDFT	NLEB = K NRAD = L	Use a smaller grid in TDDFT than in DFT.

[a]Requires LAPACK (at present, only implemented for MKL).
[b]MODPAR and NGUESS are packed options, and the values here should be added to other options. In practice, MODPAR = 73 (or 65) and NGUESS = 10 can be used.
[c]For a positive value K, K sets of integral+label are stored (for a negative K, $|K|$ words of memory are used). It is recommended to set NAODIR when NINTIC is used.

$FMOPRP can also be set: for example, NAODIR = 150 means that any fragment with 150 AOs or less is processed with the NINTIC feature; larger fragments are computed with direct SCF. Because the number of integrals to be stored (excluding those sieved out with the screening) depends on the molecular geometry and basis set, it is not easy to choose NAODIR.

For DFTB, MODGAM has a major impact on performance. It affects the computation of electrostatics, via pretabulating the values of γ and Γ functions. MODGAM = 13 is a good way of accelerating FMO-DFTB, although it requires a quadratic memory in terms of the number of atoms. MODESD = 2 computes separated RESDIM dimers in a compact summed form, which is not implemented for PBC. MODESD = 2 should not be used for analyses, such as PIEDA or PA.

Another acceleration route is via reducing memory (Section 5.9), because as a side effect, for some options there is also an acceleration boost. The downside of this route is that some properties are added up on the fly and individual contributions are not available in the final summary.

The idea of setting NLEB and NRAD in $TDDFT is to use a different, less accurate grid in TDDFT as compared to underlying DFT. The values of these options in $TDDFT can be set depending on similar options in DFT. The default setting is $DFT NRAD = 96 NLEB = 302, and for accelerating TDDFT, $TDDFT NRAD = 48 NLEB = 110 $END are set by default (only certain "quantized" values of NRAD and NLEB listed in docs-input.txt are allowed).

5.8.3 *Acceleration via Scientific Means*

Some acceleration can be achieved by changing scientific methods, as summarized in Table 80. Using some of these options such as RIMP2 changes the results (RI-MP2 vs. MP2).

It is possible to accelerate the calculation of ES dimers via the use of the multipole expansion, which is a fast way of computing the energy of far separated dimers (not available in DFTB). The expansion can be turned on with MODMUL = 1 (compute individual dimer values) or 2 (compute only the total sum). The latter is faster, but it hides the values

Table 80. Ways to achieve acceleration by changing scientific details.

Group	Option	Meaning
$FMO	MODMOL = K^a	Skip some dimer calculations.
$FMO	MODMUL = L^b	Use multipoles for ES dimers.
$FMO	IEXCIT(3) = 0	Skip ground state dimer/trimers for RUNTYP = ENERGY in TDDFT.
$MP2	CODE = RIMP2	Use RI in MP2.

[a]K = 1 or 3, see Section 2.6.2.7.
[b]L is 1 (analyses) or 2 (optimization or MD).

of individual PIEs. This option is especially efficient when there are many small fragments, for example, in a large cluster of water molecules.

5.9 How to Calculate Millions of Atoms

Most FMO calculations have no limit to the number of atoms. It is possible to do geometry optimizations and MD simulations for systems consisting of millions of atoms. The current record is slightly over 1 million atoms for both RUNTYP = OPTFMO and RUNTYP = MD, but the size can be pushed further. There are two aspects of running million jobs: fitting the job in memory and the sheer horsepower (and, possibly, network performance). In practice, because of the computational cost, all million jobs so far have been done with DFTB; for other methods the record is a few dozens of thousands of atoms.

It is impossible to calculate millions of atoms without using special tricks described here. These tricks are tricky to use (tautological as it may sound), which can be tried by veteran expert users. These special features are subject to various restrictions with respect to combining them with other options.

Not all calculations can be practically conducted for millions of atoms: analytic Hessians, PCM, PIEDA, and LCMO are examples of jobs that are not feasible for this size, as limited by memory. Some calculation types are limited by MXATM = 2000 atoms, for example, RUNTYP = OPTIMIZE or SADPOINT. What can be run for millions of atoms is a DFTB calculation without implicit solvent and PBC, excluding analytic Hessians (semi-numerical PAVE may be possible).

Whether or not supercomputers can be used to perform non-DFTB million jobs remains to be seen. On the other hand, million jobs with DFTB can be done on a single PC node, although the horsepower aspect is the limiting factor for long geometry optimizations and MD trajectories. The reported million job of a 10 μm white graphene nanomaterial with FMO-DFTB was a demonstration of MD performed for 500 fs, in which all atoms were treated quantum-mechanically (see Section 6.6). Likewise, a geometry optimization of a nanocrystal of fullerite with over 1 million atoms was conducted with FMO-DFTB. Both simulations used a few common PC nodes.

To print a list of pair interactions at the end, FMO2 has to store pairwise information in memory, resulting in several $O(N^2)$ arrays, where N is the number of fragments. Likewise, in FMO3, there are $O(N^3)$ arrays. However, it is possible to reduce the scaling of the required memory by 1 order, that is, to $O(N^1)$ for FMO2 and to $O(N^2)$ for FMO3 (subject to some conditions). Such calculations produce correct total properties, but without a list of MBIEs (individual values are still printed, scattered in output files). These techniques are most useful for geometry optimizations and MD.

For FMO3, many of the FMO2 memory reduction techniques may not be used, but there are separate FMO3-specific tricks. However, million jobs are only feasible with FMO2. Reducing memory in many cases has a side effect of accelerating calculations, because the code in some critical places is written differently for these memory-sparing options. Some acceleration options, however, take too much memory and have to be disabled, such as MODGAM = 13 in $DFTB.

The options useful for large jobs are listed in Table 81. To reduce the memory scaling of FMO3, MODPAR = 32 and NBUFF options can be used (NBUFF should be set to the number of SCF trimers, or an upper estimate thereof). For FMO2, MAXRIJ can be set to the maximum number of SCF dimers per fragment, so that the size of the distance array can be reduced.

For large jobs, printing atomic charges should either be disabled (NPRINT = 8 in $FMOPRP should not be used), or MODCHA = 1 in $FMO should be set.

FMO1 calculations are rarely used, so they are not explicitly described here. As a general guidance, acceleration options for FMO2 may also be set for FMO1.

For homogeneous molecular clusters, it is faster to use the NACUT option compared to an explicit definition of fragments in INDAT.

The header for a real example of a million job (Section 6.6) is shown in Table 82. In this BN nanosystem there are many detached bonds, and the job type is MD. In the input, instead of using the superoption MODIO = 3072, some relevant options are set by hand.

MODIO = 543 is 512+16+8+4+2+1, of which 31 is to reduce I/O and 512 is to use an XYZ file instead of an $FMOXYZ group. The file has to be copied to the run-time directory of every node before

Table 81. Options to reduce memory requirements and accelerate[a] FMO*n* for large jobs.

n	Group	Option	Effect	Requirement
2	$FMOPRP	NPRINT = 131	Reduce print-out.	
Any	$FMO	MAXRIJ = K[b]	Remove quadratic distance array.	RESPPC = RESDIM = RCORSD.
Any	$OPTFMO	METHOD = CG	Remove quadratic inverse Hessian matrix.	
Any	$SYSTEM	Add 512 to MODIO	Use external XYZ file for atomic coordinates.	XYZ file should be copied to the run-time directory SCR.
2,3	$FMOPRP	Add 32 to MODPAR	Remove dimer ($n = 2$) or trimer ($n = 3$) arrays.	No PCM, AP, MFMO, or PIEDA.
3	$FMOPRP	NBUFF = L[c]	Remove some trimer arrays.	1 should be added to MODPAR.
2,3	$DFTB	MODESD = 16	Accelerate MODESD = 2.	DFTB with MAXRIJ.
2,3	$DFTB	MODGAM = 0	Remove γ and Γ arrays.	DFTB.

[a]The acceleration options in this Table are for huge jobs; for general ways see Section 5.8.
[b]K should be set as explained in the main text, usually, $K = 5...10$.
[c]L should be set as explained in the main text, usually, $L = 15N...30N$, where N is the number of fragments..

running the job. KDIAG = 5 chooses the LAPACK matrix diagonalization, for which MKL should be used in building GAMESS. MEM10 is set to reduce I/O by placing the dictionary file in memory.

NPRINT = 131 is used to reduce I/O. MODPAR = 1065 consists of 1024+32+8+1, in which 1024 is used to put F40 into memory and 32 is a major memory reduction option. NGUESS = 130 in addition to the standard 2 includes 128, a work-around suboption for a technical issue in restarting calculations during MD or optimization when in-memory F40 is used (128 disables such restarts). MODGRD = 48 is 32+16, 32 for the exact gradient with 16 added for AFO. MAXRIJ = 5 is a major memory-saving option, 5 is used here because the density of fragments (the number of fragments per volume) is relatively low.

Table 82. Header for an FMO/MD calculation of inorganic nanomaterial containing over 1 million atoms.

```
$CONTRL RUNTYP = MD NPRINT = -5 MAXIT = 60 LOCAL = POP $END
$SYSTEM MWORDS = 200 MEMDDI = 2000 MODIO = 543 KDIAG = 5 MEM10
  = 20000000
$END
$FMOPRP NPRINT = 131 MODORB = 3 MODPAR = 1065 NGUESS = 130 $END
$BASIS GBASIS = DFTB $END
$DFTB SCC = .TRUE. DISP = UFF MODGAM = 0 MODESD = 18 PARAM =
  MATSCI03 $END
$MD
   MDINT = VVERLET DT = 1.0D-15 NVTNH = 2 NSTEPS = 1000
   MBT = .TRUE. MBR = .TRUE. BATHT = 298.0 RSTEMP = .TRUE.
   JEVERY = 1 KEVERY = 10
$END
$FMO
    NLAYER = 1 NBODY = 2 NFRAG = 65550 MODGRD = 48 MAXBND = 131101
    RAFO(1) = 0.1, 1.1, 1.1
    MAXRIJ = 5 MODESP = 16384
...
$END
```

MODESP = 16384 is a way to truncate electrostatics for really huge systems in DFTB.

MAXBND is set to the number of detached bonds plus 1 (by defining the exact number, the required memory is slightly reduced). MODESD = 2+16 is a fast way to compute ES dimers in DFTB. UFF is a faster dispersion model than D3. MODGAM = 0 is used willy-nilly, because the fast MODGAM = 13 option requires a memory quadratic in the number of atoms (if your computer happens to have $\sim(10^6)^2/2$ words (4 TB) of RAM per core, it will be very handy here).

Because GUI are rather overwhelmed by such molecular sizes (VMD being a laudable exception), the input file had to be made with an aid of scripts, that generated INDAT, $FMOBND, etc.

A summary of this calculation is shown in Table 83. Admittedly, on a single PC node the calculation is relatively slow and a long MD may be hard to perform. In two separate runs, 40 and 80 compute

Table 83. Summary for an MD of a BN nanomaterial containing over 1 million atoms.

Property	Details
Computer hardware	1 node, dual 20-core 3.1 GHz Xeon (40 cores total)
GAMESS model	ifort/MKL/sockets
Longest linear dimension	10.7 µm
Atoms	1180800
Basis functions	3542400
Fragments	65550
Detached bonds	131100
Used replicated memory (MWORD) per core	149
Used shared memory (MEMDDI) per node	1274
Input file size (no coordinates inside)	8.4 MB
XYZ file (coordinates)	56.3 MB
Timing per MD step, min	11

processes were executed although only 40 physical CPU cores are present in the node. With 40 and 80 processes the timings were 14 and 11 min, respectively, thus there is some benefit of hyperthreading but not 100%. The use of 80 processes for an unknown reason (probably, some OS limit) would not start with NGROUP = 80 (80 logical nodes with 1 core), but the calculation worked with NGROUP = 1 NSUBGR = −1 in $GDDI (1 node split internally).

5.10 Improving GDDI Performance

Managing a large crowd of human workers and dividing work among them is a major challenge. They can waste too much time talking to each other and in fact having too many workers on a small project can slow down the progress — the job may be done faster by reducing the team. It is the same with CPU cores. It is difficult to get good performance if there are too many cores, and one can sometimes get the job done faster by reducing the core count. A good management is the key to success with CPU cores too. One cannot just run a GAMESS job blindly and expect it to work marvels. A part of management of parallelization is left with the user (you!) in the form of input options.

FMO can deliver high parallel performance for many cores. However, good performance does not come automatically. The user has to apply some "levers" to help the program.

5.10.1 *Analyzing Parallel Performance*

Before learning to use advanced options, it is necessary to understand how to get timing data. GAMESS prints separate CPU and wall-clock times. A wall-clock time is the elapsed astronomical time (what if GAMESS runs on a spaceship travelling near the speed of light?). A CPU time shows how much CPU actually worked. The difference between these two timings is typically due to CPU slacking off during I/O, waiting for data access, or in a synchronization.

An example is shown in Table 84. For most users, only the total CPU timings are of interest (447.8, 0.3, 0.3, etc.), and the wall-clock time (463.0) is always the same on all ranks because of synchronizations. It can be seen that something is very wrong in this calculation: the CPU time on rank 1 is 448 s, and on other ranks, 0.3 s, showing

Table 84.   **Example of a summary of CPU and wall-clock timings (s).**

	Rank[a]	total	SCF	corr	DFT	CPHF	EFP	solv	ESP	geom
		Total timing statistics in pairs (CPU,wall) for each rank, s								
CPU	1	**447.8**	432.8	0.0	0.0	0.0	0.0	0.0	14.5	0.0
wall	1	463.0	433.4	0.0	0.0	0.0	0.0	0.0	14.5	0.0
CPU	2	**0.3**	0.0	0.0	0.0	0.0	0.0	0.0	0.0	0.0
wall	2	463.0	0.0	0.0	0.0	0.0	0.0	0.0	0.0	0.0
CPU	3	**0.3**	0.0	0.0	0.0	0.0	0.0	0.0	0.0	0.0
wall	3	463.0	0.0	0.0	0.0	0.0	0.0	0.0	0.0	0.0
	...									

[a]A pedantic reader may notice that the number printed here is the rank plus 1.

that the load balancing is in a state of disaster of a legendary scale (discussed more in Section 5.10.4.2).

I/O can be reduced by using fileless options (Section 5.8.1), although some methods use disk anyway, such as IMS MP2. Another route is to change the method, for example, switch from CODE = IMS to DDI in MP2 (Section 4.3.2). Direct SCF (DIRSCF = .T.) can be used, and there are other ways to reduce I/O (Section 5.8.1). A synchronization problem can arise because of a poor load balancing (Section 2.9.3).

Using timings in Table 84, CPU utilization ratios for each rank can be computed (dividing CPU by wall-clock timings). Ideally, they should be 1 (or 100%). Two cases can be encountered: (A) all ratios are roughly similar, and (B) some ratios are high (diligent workers) and some are low (idle loafers). Four common classes of issues are summarized in Table 85.

If CPU ratios of all groups are close to 100%, then congratulations, this run is very good and nothing needs to be changed. If, however, all ratios are much less than 100%, it is likely that (A1) there is an I/O problem, or (A2) a network bottleneck. For A1, the solution is to reduce I/O. For A2, one solution is to reduce network overhead by using only 1 node per group. Distinguishing A1 from A2 takes some experience; as a sign, A1 can arise even on a single core, whereas A2 occurs only if multiple cores are used.

An A2 case of a network topology problem is illustrated in Figure 5. When CPU cores are divided into groups, the details of the number of cores per CPU, network connectivity, etc., should be taken into account for good performance, which may require assigning core affinity and/or using a manual setting of group sizes (MANNOD). Some queuing systems may be prone to giving a scattered list of nodes, because they do not know about groups (and why should they).

An especially troublesome case of (A) is when multiple cores get in each other's way, for example, by using too much CPU cache. In this case, it is neither a true network overhead problem, nor an I/O problem. This problem in practice may be limited to DFTB, with a solution to reduce the core count per node.

Table 85.   Causes and solutions for poor performance.

Label	CPU utilization ratio	Cause	Cure
A1	Low on all groups.	Poor I/O	1. Set DIRSCF = .T. 2. Define SCR for run-time files on a local disk. 3. Especially for DFTB: use MODIO = 3072. 4. Use an equivalent method with a lighter I/O overhead (for MP2, CODE = DDI).
A2	Low on all groups.	Network efficiency	1. For DFTB: use 1 node. 2. Use an equivalent method with a lighter parallel overhead (for MP2, CODE = IMS). 3. Define groups in agreement with the network topology (use MANNOD, tweak MPI, set core affinity).
B1	Low on some groups for monomers.	Load balancing	1. Use semi-dynamic load balancing for monomers. 2. Set NGRFMO(i) for i = 1, 4, or 6.
B2	Low on some groups for dimers/trimers.	Load balancing	1. Use semi-dynamic load balancing for dimers/trimers. 2. Set NGRFMO(i), i = 2 or 7 (dimers) or NGRFMO(3) for trimers.

A radical way to deal with A2 is to upgrade network, such as to replace Gigabit (whose weak point is latency) by Infiniband. The effect can be from nearly zero to a large boost, very much depending on the details of the run. In many cases the bottleneck of an FMO parallelization is load balancing, for which upgrading network can have no noticeable effect.

The B group of issues is related to load balancing. B1 can occur when a fragment takes much more time than the rest, for example, its convergence is very poor, or its QM method is expensive (MCSCF or TDDFT).

B2 is similar to B1, but in reference to dimers and trimers. Because there are more dimers than monomers, the load balancing for dimers is usually better (more tasks are easier to divide). In addition, in the monomer loop there are multiple synchronizations (pertinent to B1). To decrease synchronization loss, one can try to reduce the number of dimer groups or use the semi-dynamic load balancing.

5.10.2 *Toll of Data Servers*

It is useful to check how much time data servers use (note that on 1 logical node with socket DDI, data servers are not employed). The moment of truth has come for a close look at skeletons in the DDI closet.

A sample of the timing output is in Table 86. Here, 40 logical nodes of 1 core were used (rungms job 00 40 1 1). There are 40 compute processes with ranks 0...39 and 40 data servers with ranks 40...79.

The results uncover that most data servers indeed did not use CPU (this was a socket DDI version), but the grandmaster data server (rank 40) had a CPU usage of 8.8 seconds. It has to work more than other data servers (like a good manager), because it has to issue

Table 86. Sample of timings for 40 compute processes and 40 data servers.

CPU timing information for all processes	Comments:
0: 41.221 + 2.497 = 43.719	grandmaster compute process 0
1: 37.495 + 1.639 = 39.135	
2: 37.864 + 1.874 = 39.738	
3: 37.463 + 1.989 = 39.453	
4: 37.630 + 1.687 = 39.317	compute process 4
...	
40: 0.806 + 8.33 = 8.839	grandmaster data server 0
41: 0.02 + 0.02 = 0.04	
42: 0.01 + 0.02 = 0.04	
43: 0.00 + 0.03 = 0.04	
44: 0.00 + 0.06 = 0.06	data server 4
...	

counters for dynamic load balancing. The two timings printed for each rank are so-called utime (spent by a GAMESS process) and stime (spent by OS processes on GAMESS errands). It is curious that stime is 8.3 s on the grandmaster data server.

5.10.3 *Three-Level GDDI*

In this book, two-level GDDI/2 is mainly described, which is the workhorse parallelization engine of FMO. There is, however, a more advanced GDDI, in general denoted by GDDI/*n*, which in practice is limited to GDDI/3, a three-level GDDI (Figure 26).

GDDI/3 can be thought of as a set of multiple concurrent instances of GDDI/2. In GDDI/3, the whole set of CPU cores, called the universe, is divided into a set of worlds. Each world is of a GDDI/2 type, consisting of a set of groups, and each group is a set of processes. Thus, there are three levels of parallelization: dividing the workload between worlds, groups, and processes.

To use this level of complexity, the computational task at hand must be structured in a proper way. At present, only two tasks can be run with GDDI/3: RUNTYP = HESSIAN (semi-numerical Hessians) and RUNTYP = MEX (finding a minimum energy crossing of two spin surfaces). For MEX the number of worlds is fixed to 2, one world per multiplicity. MEX is special, because without GDDI/3, MEX cannot

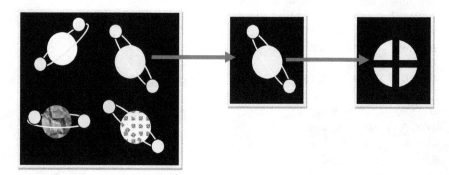

Figure 26. Schematic structure of GDDI/3. The universe is composed of 4 worlds, each world has 3 groups, and each group is divided into 4 members.

preserve fragment data for each multiplicity separately. Therefore, for MEX the use of GDDI/3 is vital.

For semi-numerical RUNTYP = HESSIAN, the number of the worlds is arbitrary and can be chosen by the user. For DFTB, a group is limited to 1 process, and GDDI/3 framework can be used for an effective two-level parallelization at the level of worlds and groups.

GDDI/3 is set up by defining NGROUP = K worlds in $GDDI, for example, NGROUP = 2 for MEX. The number of groups per world is set with NSUBGR. Many advanced manual options in GDDI/2 such as MANNOD are of limited use in GDDI/3.

NSUBGR = −1 has a different meaning not related to GDDI/3; this setting is used for dividing intranode cores into teams, one core per team, so that NSUBGR = −1 is effectively a special one-level parallelization. NSUBGR = −1 is mainly used for sockets (although it can be used for MPI).

In MPI, there is a better, more general way of doing intranode parallelization, accomplished by providing the 5th argument to rungms (for sockets this 5th argument is only used for a single node). This argument sets up the environmental variable DDI_LOGICAL_NODE_SIZE. The value of this variable set to 1 is roughly equivalent to NSUBGR = −1. For example, when running MPI on 3 nodes with 8 compute processes each, setting DDI_LOGICAL_NODE_SIZE to 2 will result in $3 \times 8/2 = 12$ logical nodes of 2 compute processes each, and GDDI can use up to 12 groups, whereas without the variable the maximum number of groups is 3.

5.10.4 *How to Optimize Parallel Efficiency*

It is instructive to consider several representative test cases and highlight the problematic points in the parallelization. All tests in this section are performed on a single node with a dual 20-core Xeon CPU (40 cores total). In terms of arguments to rungms, the simplest is to split a single node with 40 cores into 40 logical nodes of 1 core each when running rungms, accomplished with:

rungms job 00 40 1 1

With this way of running GAMESS, it is possible to use from 1 to 40 groups in GDDI. A marginally better performance can be achieved by finetuning arguments of rungms to the value of NGRFMO, for example, to use NGROUP = 1, the unsplit node can be used with

rungms job 00 40

There are two causes of bad parallel efficiency: communication cost and synchronization loss. The former can be improved by increasing the number of groups while decreasing the group size, and for the synchronization, by the opposite. The balance between the two causes is the key toward getting good parallel performance. Communication cost can be decreased by investing into better networking hardware, but synchronization loss can only be improved via algorithmic means. Think of a train crossing: a snail and a sports car spend the same time waiting. Having a fancy networking may not improve performance if the bottleneck is synchronization.

Tests here are limited to FMO2; by extension, similar techniques can be applied to FMO3. In all tests a single physical node is used. Having more nodes may permit discussion of additional fine points, but the main issues of load balancing can be perfectly well learned on 1 node. On massively parallel computers, when the number of cores is at least a 3-digit number, peculiar problems may arise that are not discussed here.

5.10.4.1 *Similar fragment sizes*

A water cluster $(H_2O)_{256}$ is calculated at the level of FMO2-RHF/6-31G* in two ways: using 1 and 40 groups. This test serves as an illustration against a black-box usage trusting the program *to sort it out somehow*. The timings for NGROUP = 1 and NGROUP = 40 are 433 and 31 seconds, respectively. Thus, a speed-up of 14X can be achieved by changing one keyword NGRFMO. It may be noted that the speed-up is relatively small, because communications within a node are fast owing to shared memory. If multiple nodes are used, the speedup can easily be a 3-digit number.

Table 87.   Timing statistics for (H₂O)₂₅₆ run on 1 group with 40 cores.

	Total timing, s	
	Rank	**Total**[a]
CPU	1	185.2
wall	1	433.0
CPU	2	136.8
wall	2	433.0
	...	
CPU	40	86.3
wall	40	433.0
CPU	averaged	**127.3**
wall	averaged	**433.0**

[a]Other columns in the output Table are truncated.

How can the user have suspected that "something is rotten in the state of Denmark", just from the results with 1 group? Total timing statistics (Table 87) indicate that the averaged CPU time is 127 seconds compared to the wall-clock time of 433 seconds. This shows a low CPU usage (29%). Looking at the timings on various nodes, one can see that all of them are quite bad, whereas the fragment size is uniform (one water per fragment).

As a summary, the cause for poor performance is intragroup communications and synchronization; a solution is to use 40 groups. The take-home lesson here is, if there are many more fragments of similar size than cores, assigning 1 core per group is efficient.

5.10.4.2 *Uneven fragment sizes*

Using many groups can decrease performance. Let us consider an organic molecule ibuprofen (33 atoms) with 9 water molecules calculated at the level of FMO2-RHF/6-31G*.

The number of atoms in ibuprofen (33) is 11 times larger than in water (3), and the number of AOs (261 vs. 19) is ~14 times larger. The cost of QM calculations of a single fragment is at least cubic in terms of the number of AOs, with some steps scaling as the fourth power (integrals), the fifth power (MO transformation in MP2), and higher powers (CC). This should be remembered when designing load balancing. Assuming the lowest cubic power, the cost of calculating ibuprofen is $14^3 \approx 2500$ times more than that of water. Perhaps now the difficulties in doing load balancing start to daunt.

Four calculations were performed using M^{mon} groups for monomers and M^{dim} groups for dimers, summarized in Table 88. The reasoning for $(M^{mon}, M^{dim}) = (1,9)$ is the number of large tasks. There is 1 big ibuprofen monomer, hence 1 monomer group; there are 9 big ibuprofen-water dimers, so 9 dimer groups. This choice shows a good skill of the user. The setup (10,13) is a semidynamic load balancing, devised by the expert user. 10 groups are set up for monomers: 1 group with 31 cores (for ibuprofen), and 9 groups with 1 core (for water). For dimers, 9 groups are set with 4 cores (ibuprofen-water), and 4 groups with 1 core (water-water). The choice of 10 and 13 groups is not unique, just one possibility among many.

Analyzing the timings, the first choice of 1 monomer and 1 dimer groups has a decent performance, because only one node is used so communications are fast. The timings are dominated by the few large

Table 88.   Wall-clock timings T (s) for an organic molecule ibuprofen with 9 water molecules, using M^{mon} groups for monomers and M^{dim} groups for dimers.

M^{mon}	M^{dim}	T	Input options	Reason
1	1	67	NGROUP = 1	The simplest choice.
40	40	463	NGROUP = 40	Many groups are good.
1	9	56	NGRFMO(1) = 1,9	Set the number of groups to the number of large tasks.
10	13	55	NGRFMO(1) = 10,13 MANNOD(1) =	A semi-dynamic load balancing:
			31,1,1,1,1, 1,1,1,1,1,	10 groups for monomers,
			4,4,4,4,4, 4,4,4,4,	13 groups for dimers.
			1,1,1	
			LOADBF(1) = 260,279[a]	AO thresholds: mon.,dimers.
			LOADGR(1) = 1,9	Groups for static load balancing.[b]

[a]Chosen so because there are 261 and 280 AOs in ibuprofen and ibuprofen-water dimers, respectively.
[b]Using 1 group of 31 cores for large monomers, and 9 groups of 4 cores for large dimers.

jobs, in which case few groups work well. The choice of (40,40) is an epic blunder, and in the author's experience, an extremely common one too. 10 fragment tasks are distributed among 40 groups, so 10 cores work and 30 cores do nothing but consume resources (not unlike a human organization).

The other two choices are made by the competent user. The relatively simply choice (1,9), where some thought is infused, works very well. A negligible improvement is obtained by a far more complicated choice (10,13). Comparing 56 and 463 seconds for (1,9) and (40,40), a speed-up of 8X is achieved.

CPU timings for the case of (40,40) groups are shown in Table 89, whereas the wall-clock time is the same (463 s) for all ranks due to synchronizations. Rank 0 worked 448 s, because it did ibuprofen (33 atoms) calculations 8 times (in the monomer loop), and also one ibuprofen-water dimer. Eight other ranks worked 30–90 s, doing one ibuprofen-dimer each, and some water. There is a range of timings (30–90 s) even though the task size is the same, as driven by the number of SCF iterations to converge and the density of nearby fragments for the ESP calculation. Sinecure groups that did only water monomers and water-water dimers worked for just 2 seconds at most, and some as little as 0.3 s.

How can the user get a scent that something is rotten in the electronic state of FMO? Perhaps the easiest is to use "grep synch" on the output file of each group (Table 90). It shows that group 0 did not wait (0 s) in the monomer calculation, because it had the largest task to do (ibuprofen) and all other ranks doing water waited for it (see

Table 89. **CPU timings (s) on each rank for a (40,40) group calculation of solvated ibuprofen.[a]**

Ranks	Timings
0...9	447.8 0.3 0.3 0.5 0.7 0.9 0.9 1.2 1.2 71.3
10...19	32.0 70.1 84.7 87.0 80.9 79.3 87.3 2.0 1.7 1.7
20...29	1.6 1.3 1.3 1.1 1.1 1.1 1.1 0.3 0.3 0.3
30...39	0.3 0.3 0.3 0.3 0.3 0.3 0.1 0.2 0.3 0.3

[a]This spectacular run is also featured in Table 84.

Figure 7 for a visualization of idle waiting at a synchronization barrier), for example, rank 1 waited between 16 and 103 s. The waiting tends to decrease as the electronic state converges in the monomer loop. For dimers, rank 0 waited for 14 s. This is because it finished its large ibuprofen-water dimer, and waited for other groups doing other similar dimers, some of which took more SCF iterations. Rank 1 did not get any ibuprofen-water dimer, and only quickly did its share of water-water dimers, and then it pathetically waited for 86 s.

In this example, one fragment is large in size. What really matters for load balancing is the cost, which is determined by (a) size, (b) wave function, (c) SCF convergence, and (d) environment (how close other fragments are). In some FMO methods, like MCSCF, UHF, or TDDFT, different wave functions are used in the same run. For example, in MCSCF, some fragments are computed with RHF. In this case, the user should consider not just the size but also how much

Table 90. Synchronization timings in two groups for solvated ibuprofen with the (40,40) group definition.

Group 0		
Monomer SCF synchronization on iteration	1.1 took	0.0 s.
Monomer SCF synchronization on iteration	2.1 took	0.0 s.
Monomer SCF synchronization on iteration	3.1 took	0.0 s.
Monomer SCF synchronization on iteration	4.1 took	0.0 s.
Monomer SCF synchronization on iteration	5.1 took	0.0 s.
Monomer SCF synchronization on iteration	6.1 took	0.0 s.
Monomer SCF synchronization on iteration	7.1 took	0.0 s.
Monomer SCF synchronization on iteration	8.1 took	0.0 s.
Dimer synchronization for layer 1 took	14.4 s.	
Group 1		
Monomer SCF synchronization on iteration	1.1 took	102.7 s.
Monomer SCF synchronization on iteration	2.1 took	71.3 s.
Monomer SCF synchronization on iteration	3.1 took	60.5 s.
Monomer SCF synchronization on iteration	4.1 took	46.0 s.
Monomer SCF synchronization on iteration	5.1 took	36.3 s.
Monomer SCF synchronization on iteration	6.1 took	25.5 s.
Monomer SCF synchronization on iteration	7.1 took	17.5 s.
Monomer SCF synchronization on iteration	8.1 took	16.3 s.
Dimer synchronization for layer 1 took	86.3 s.	

more expensive it is to run MCSCF compared to RHF. Some electronic states, for example, of iron-containing fragments are difficult
to converge and take many iterations, not directly related to the fragment size.

In this test, the cause for poor performance is synchronization,
and a solution is to use 1 group or semi-dynamic load balancing. The
take-home lesson here is, if one fragment is costly and the rest are
fast, set the number of groups to the number of large tasks (the
responsibility to count large tasks lies with the user: no brain pain,
no gain).

5.10.4.3 *Multiple layers*

A common mistake in executing an FD calculation (Section 4.7) is
the failure to take into account that the number of tasks in layers
differs (if domain **F** is not empty). As an example, phenol with $(H_2O)_8$
is divided into 9 fragments. As discussed in Section 4.6, all $N = 9$
fragments are computed in layer 1. Domain **A** includes only phenol,
and **b** has 2 extra water molecules, so $B = A \cup b$ (layer 2) has phenol
and 2 water fragments. Three GDDI setups are attempted (Table 91),
denoted by $N_1^{mon}, N_2^{mon}, N_2^{dim}$, where N_L^A is the number of groups used
for A = mon (monomers) or dim (dimers) of layer L (in layer 1, dimers
are not computed in FD).

The attempted 3 ways of defining groups are 9,9,9 (because there
are 9 fragments), 1,1,1 (because phenol is larger than water and the
number of cores is relatively small), and 1,1,2 groups (after analyzing
the failures of the other runs). Specifying K, L, M groups for two layers
in FD is achieved by NGRFMO(1) = K,0,0,0,0, 0,0,0,0,0, L,M (further
trailing zeros may be omitted). The level of calculations is FMO2-
RHF/6-31G*:B3LYP/6-31G**.

The FD peculiarity is that for the initial geometry only (with the
timing of T_1), monomer calculations are done for the first layer, followed by monomer and dimer calculations of the second layer,
whereas for all further points (T_2), only the second layer is computed.
The use of 9 groups is not fast (29.4 s), because 9 fragments have
uneven sizes. Although for the best performance one could design

Table 91. Timings (s) for 2 geometry optimization steps (T_1, T_2) in FMO/FDD for phenol+$(H_2O)_8$, on 1 node with 40 cores, divided in N_L^A groups (A = mon or dim) in layer L.

N_1^{mon}	N_2^{mon}	N_2^{dim}	T_1	T_2
9	9	9	29.4	18.0
1	1	1	17.5	8.9
1	1	2	16.2	7.8

customized groups, the use of 1 group is a simple and useful choice in this system when run on 1 node (17.5 s).

For the second layer, the simplest solution $N_2^{mon}, N_2^{dim} = 1,1$ is not the best. Using 9,9 is clearly a blunder because there are only 3 fragments in layer 2 assigned to 9 groups (think of asking 9 postmen to deliver 3 letters). A good solution is to use 1 group for monomers (one large solute) and 2 for dimers (2 large solute-water dimers), with the timing of 16.2 s. The overall difference between the best and the worst is ~2X. It is relatively small because fragments in this calculation are too small for 40 cores, so using few large groups has a substantial penalty.

The take-home lesson here is, when choosing the number of groups, think of the number of tasks; if tasks are of uneven size, focus on the large ones.

5.10.4.4 *Protein-ligand complex*

The final example is meant to be a fairly common calculation, different from the extreme cases above. It is a protein-ligand complex (1L2Y with *p*-phenolic acid), calculated as PIEDA MP2/PCM/6-31G*. The idea is to start with a guess for group sizes, analyze the results and devise an improvement. In many real-life applications, the number of cores is much smaller than the number of fragments, so the effect of choosing a group size is minor, and the golden rule (Section 2.9.1) works well.

There are 21 fragments in this complex. A guess was made that the scaling factor a_2 for SCF dimers in the golden rule is equal to 3, which is often seen in linear chains (an FMO calculation prints the total number of SCF dimers (N_2) and trimers (N_3), from which the actual a_i can be computed by dividing N_i by N for $i = 2$ or 3). In this particular system, a_2 is 2.9.

The golden rule for $N = 21$ says to use NGRFMO(1) = 7,21 (there are no trimers). The timing obtained with the golden rule is 39 min (on 40 cores) with the averaged CPU utilization of 94.9%. It is already good enough, but one attempt is made to improve it. It was found by doing a grep for "synch" that dimer synchronization is about 1 minute. This may be caused by having groups of different sizes (for 40 cores divided into 21 groups, some groups have 1 and some 2 cores). Thus, NGRFMO(1) = 7,10 was attempted. It decreased the dimer synchronization to 2 seconds. The timing was 38 min, thus the goal of cutting the waste of 1 minute was attained.

The take-home lesson here is, the golden rule does work (most of the time).

5.10.5 *File-Less GAMESS Execution*

The modern paradigm for computational efficiency has been pushing program developers to avoid I/O. It may be impossible to avoid using files altogether, because at least input data and output results should be handled, but there are quite a lot of other files that can be got rid of. The default way of running GAMESS results in writing the main output file (the log file) on a network disk. At present, no user-controlled way exists for redirecting the output to a local file, although it is easy to accomplish via a small change in the program.

In Section 5.8.1, I/O reduction is described in some detail. Here, a summary of some ways to reduce file usage as pertinent to FMO is described in Table 92 as a convenient reference. Many of these options apply to runs without FMO as well. In GDDI, many of the listed files are created on each rank, or in each group, so the total number of files can be overwhelming. In most cases there is an acceleration due to reducing I/O. In some cases, as for DIRSCF, it may actually be faster to write data on disk. For F08, two ways are listed, with a different strategy for removing the file.

Table 92.   Options to avoid using files in GAMESS.[a]

File	When used	Options to accomplish
F06	FMO output in GDDI (non-master)	$GDDI PAROUT = .F. $END
F06	FMO output in GDDI (all)	$SYSTEM MODIO = 4096 $END
F07	FMO DAT in GDDI (non-master)	$GDDI PAROUT = .F. $END
F08	SCF	$SCF DIRSCF = .T. $END
F08	SCF	$INTGRL NINITC = K $END
F09	MP2	$TRANS DIRTRF = .T. $END
F10	Always	$SYSTEM MEM10 = K $END
F15	SOSCF for HF	$SYSTEM MODIO = 32 $END
F22	DFT	$SYSTEM MEM22 = K $END
F23	FMO-DFTB restart	$OPTFMO MAXNAT = K $END
F40	FMO fragment data	$FMOPRP MODPAR = 1024 $END

[a]Suitable values of K are discussed in this book. For packed options such as MODIO and MODPAR, the values given here should be added to other suboptions.

CHAPTER 6

Reference Materials

6.1 Conclusions and Outlook

Various ways of executing FMO calculations in GAMESS have been described. This book is a practical field guide and is meant to be not just read, but the reader is expected to execute various calculations and sharpen the skills of using FMO in GAMESS.

Owing to the excellent QM engines in GAMESS that FMO heavily uses, it is possible to do a wide variety of applications in the fields of physical chemistry, biochemistry, solution chemistry, photochemistry, their sister areas in physics, and material science. The latter is a nascent application field for FMO owing to the recent development of methods with periodic boundary conditions.

FMO can generate all kinds of properties for data mining and machine learning, to be used in designing new materials.

The ability to use multiple CPU cores and nodes makes it possible to complete projects involving large systems in a realistic time. However, parallelization is not a black box. To gain good performance, the user should get some proficiency in the use of the fairly complex machinery of the multi-level parallelization.

The author hopes that this book can explain many difficult-to-understand aspects of running FMO in GAMESS by focusing on the important options and omitting the unimportant. The number of input options is colossal and their combinations astronomical, but among those, only relatively few are of importance. They are described in this book.

6.2 Troubleshooting

So you installed GAMESS, made an input and executed it. But it did not work. It happens so often to everyone, including program developers. What to do next? Although the full variety of problems is impossible to cover, here some common causes are explained.

One comment should be made, if anything else, for attaining the peace of mind of the user. Many error messages give the immediate cause of the problem and do not suggest a solution. Imagine you fell off a cliff. Your mind reports that you cannot continue because your eyeglasses are broken; this is your error message. It is critical if your eyesight is poor and you cannot walk further without glasses. It may be that you also have broken bones (a much more serious problem) but as your mind does not have an X-ray, it only tells you about the glasses. It may also give you highly "helpful" information on the number of fractures in each lens, and the maximum height from which glasses may fall without breaking.

The checks of whether glasses work or not will not tell you what you should have done to avoid having your glasses broken. Maybe you fell off a helicopter, maybe some thug punched you in the face — the end result is broken glasses, and this is the error message that lenses are broken. What you should have done to avoid it is a difficult question. If it was a cliff you should have taken a different road as your navigator told you; if it was a helicopter you should have buckled up as the pilot told you; if it was a thug you should have taken karate lessons as your parent told you, etc.

The thing of primary concern for a programmer in aborting a calculation is to prevent the user from getting wrong results by continuing. If a calculation cannot recover, it is terminated. In some cases what you should have done is indeed suggested (such as try setting this option or switch to that method), and in some other cases nothing useful is suggested. It is frustrating, but you should either accept it as a fact of life or join the team of GAMESS developers and write a better program.

If the printed error message is hard to interpret, advanced users are welcome to use grep for the error message in the source code.

Often one does not even need to know FORTRAN, because there may be a helpful explanation in English (or Italian, as in the PCM code) as a comment around the error message. Perhaps a programmer wrote a comment that so many tiny pieces of glass possibly indicate falling from a helicopter, so you would know that buckling up is needed next time you fly.

The first step in dealing with a problem in running GAMESS is to identify the category it belongs to. Here, 3 main categories are described: failure to start, external abnormal termination (a GAMESS process killed by an external cause), and internal abnormal termination (the program found some conflict that cannot be resolved and aborted its execution). In addition, scientific failures and input errors are also discussed (they usually fall into the last category)

It is important to note that in case of encountering problems, it is very useful to increase the output level to gain details, where and how a calculated stopped (use NPRINT = 7 in $CONTRL, NPRINT = 8 in $FMOPRP and do not use MODIO = 3072 in $SYSTEM).

An output loss can arise due to an abnormal termination during a parallel run. Typically, only the master process writes to an output file (in GDDI, the master of each group). However, all processes execute the same code. It can happen that a compute process can reach a certain lethal condition sooner than the master, and abort the execution terminating itself first, and then in a chain all processes are slain. As a result, no error message may be printed. Sometimes OS decides to kill a GAMESS process, in which case some part of the output file not yet physically written to disk may be lost. There are some measures taken to prevent an output loss, but sometimes it happens.

If a job terminates and all GDDI output files (log and F06) have no explicit error message, and if "ABNORMAL TERMINATION" is not printed, it is likely that you are suffering an output file truncation loss. In this case, a good solution is to rerun the same job on a single core, which is very likely to produce some error message. A job may be too long to run on 1 core, and for some jobs there may be a dependence of the problem on the number of cores. However, many errors, especially those due to user errors, occur in the beginning. If

238 *Complete Guide to the Fragment Molecular Orbital Method in GAMESS*

Table 93. Typical examples of failures in running GAMESS (socket DDI).

Failure to start
(a) TCP: Connect failed. abacus1 -> abacus2:54573.
(b) poll: protocol failure in circuit setup.
(c) kex_exchange_identification: Connection closed by remote host.
ddikick.x: Timed out while waiting for DDI processes to check in.

External abnormal termination
DDI Process 12: Trapped a termination signal (SIGTERM).
ddikick.x: Application process 12 quit unexpectedly.
ddikick.x: Fatal error detected.

Internal abnormal termination
DDI Process 0 (120): error code 911
ddikick.x: Application process 120 quit unexpectedly.
ddikick.x: Fatal error detected.

not practical, the next thing would be to rerun on 1 group with multiple cores.

A sample of typical problems is given in Table 93. The issues are described in detail below.

6.2.1 *Failure to Start*

Three examples of the actual messages are in Table 93. They typically name a cause (often looking abracadabrish), followed by a timeout failure. Think of it as inviting friends to play a ball game but some (or all) did not show up (maybe they did not even get the invitation). You had honestly waited for 10 long seconds before your patience ran out and you called the game off.

A simple case of a failure to start is trying to run a socket DDI code from a node different from the master node. To use multiple nodes, rungms should be executed from the first node in the list.

Another typical cause is a network restriction (access denied). In some cases such problems can be overcome by changing the execution mode (Table 6). An absence of required shared dynamic libraries on a compute node (for example, for BLAS) may also be a problem.

Some Unix versions may limit resources: the number of semaphores and the amount of System V shared memory (shmmax) used by DDI may be limited at the OS level (see ddi/readme.ddi for details of how to modify system setup files or adjust parameters from the command prompt).

A mysteriously looking instance of a failure to run a parallel job is a lack of system memory just to load the whale pod of GAMESS executables. In other cases, network may be busy and the spawning of GAMESS processes may reach a timeout. If you kill a job, it may not release some resources, in particular semaphores. When you run another job, it can fail to start. In this case, one can use a Unix command to release stuck resources (see Section 6.4) or reboot the node.

For queuing systems, there may be restrictions on their usage (such as how many processes may be used), which are usually explained in the error message. The number of processes may be restricted by the number of physical cores, in which case only one half of cores can be used for compute processes.

Some failures to start may occur in the running script (such as mistakes in the arguments, or in the parts of rungms that deal with specific parallel environments), excruciating at times for handling different kinds of MPI libraries.

It is also possible that rungms does not support a particular queuing system, from which the script has to get some information. The socket version with an interactive execution relies on a host list specified as an argument to rungms, but for MPI procuring a host list is coded inside rungms. Some MPI versions may require setting additional variables, and for that, rungms may have to be edited.

There are limits to both the number of processes (cores) per node (MAXCPUS) and the number of nodes (MAXNODE) built into DDI (see the script ddi/compddi). An attempt to use a large cluster may end in a disappointing failure, which is, however, easy to resolve, by changing the limits, recompiling DDI (compddi) and linking GAMESS (lked).

The script rungms checks for existing restart files, and if it finds any of them in SCR or USERSCR (for instance, if an unsuspecting user is trying to run the same job again), it aborts execution with "Please save, rename, or erase...". If, however, for some reason it was spoofed off guard and a sneaky user manages to execute GAMESS while in possession of forbidden files in SCR, then an error message from GAMESS may be "ERROR OPENING NEW FILE PUNCH." The key part of it is "new" (a DAT file cannot be opened as new if it exists). Leftover DAT, restart or trajectory files with the same job name should be removed from SCR or USERSCR.

The script rungms imposes severe restrictions on the symbols that may be used in an input file name. In fact, GAMESS can process some of these symbols, but rungms is adamant in its pedanticism, allowing letters, numbers, minus, underscore, dot, and two kinds of slashes only.

6.2.2 *External Abnormal Termination*

An example of an actual message is shown in Table 93. There may be a signal termination, or a termination for overusing resources (for example, no space on hard disk). A suggestive signature of this type is an absence of an "ABNORMAL TERMINATION" in any GDDI output file. A sure sign is a termination by a signal.

Among the causes for this type of problems are: (a) bugs in GAMESS (often resulting in "KILLED BY SIGSERV"), (b) some fault with run-time files (such as a hard disk failure or running out of disk), (c) network failure, (d) tampering with the GAMESS executable while it is running (such as rebuilding it), and (e) process termination caused by a disconnection from the node for remote jobs executed interactively (why did you close that laptop lid?).

For a SIGSERV, the user can hardly do anything other than trying to report the issue, if it consistently arises. For hardware and network faults, rerunning the job may work. For resource-related issues, check the usage against hardware.

6.2.3 *Internal Abnormal Termination*

An example of a typical message is shown in Table 93. In one of the output files ($job.log or normally deleted $job.F06.*), there should be a message "ABNORMAL TERMINATION". Unfortunately, OS may lose data in buffers not yet written out to disk, so the signature message may be missing. Another sign of an internal abnormal termination is a reference to a mysterious code 911 (Table 92).

An internal abnormal termination means that GAMESS decided that it cannot continue. These terminations can be grouped into (1) input errors (mistakes in the input file), (2) unavailable methods (trying to do something that is not possible), (3) insufficient resources (usually memory), (4) divergence (in solving equations), (5) program deficiency, and (6) file error.

A file error may occur when a particular file cannot be opened, either because it does not exist, its access is restricted, or it exists and but should not. Absent parameter files in DFTB are an example.

In the case of an internal abnormal termination, an FMO dump information is written (an example is shown in Table 94). For optimizations and MD, a succinct output option (MODIO = 3072) is often used, and it may not be obvious where the program stopped without looking at the information provided by the dump. The information in the dump shows the FMO step (Table 1), and provides atomic coordinates. If the stop occurred at a fragment calculation, the coordinates of the troublemaker fragment are printed. One can visualize the fragment to see if there is anything unusual that might cause SCF divergence.

In the dump in Table 94, the calculation stopped in step 4 (Table 1), that is, in a dimer. The first line shows that the dimer is 2,1. The coordinates of the dimer are also printed. The exact reason for

Table 94. Sample FMO dump for an unsuccessful run.

FMO dump: step 4 layer 1 n-mer		2	1	0 iter 1		
NAT,ICH,MUL,NUM,NQMT,NE,NA =	6	0	1	14	14	20 10
Current n-mer						
8	2.5420270000	0.8937630000	−1.0015930000			
1	1.9918150000	1.6239620000	−1.2849790000			
1	2.9584330000	0.5812150000	−1.8048060000			
8	0.0000000000	0.0000000000	0.0000000000			
1	0.0000000000	0.0000000000	0.9572000000			
1	0.9266270000	0.0000000000	−0.2399870000			

the termination is not shown in the dump (in this particular case, SCF did not converge).

Some errors happen in steps that deal with the whole system, for example, in the initial bookkeeping (caused by user errors), optimization or MD (caused by a bad structure). One can visualize the whole system using the dumped structure to see if perhaps some undesirable deformation took place during optimization or MD. A fragment calculation can stop because SCF does not converge or due to a memory shortage.

Not all combinations of options are possible. For example, one cannot combine FMO with QuanPol. Some of these illegal combinations are explicitly trapped, and a reasonable error message is printed. However, because the number of combinations of options is exponential, it is impossible to trap all illegal combinations. One can use tea and vinegar separately, but nobody expected you combine them in one cup.

The user, if of the persistent type, by some trial and error can try to find out what is the combination of options that is not allowed, if error messages do not clearly state it, and remove non-essential options leading to the problem. Some memory problems can be resolved by an input parameter, for example, MXTS in PCM; for others, there is no way to increase the dimension other than by changing the program.

A very obnoxious (for the user) case of an internal abnormal termination is when the number of cores is too large. For example, the parallel CCSD(T) code aborts if the number of nodes is more than it can handle ("Too little work for too many nodes"); and a likewise abort can occur for other methods. A few rarely used methods cannot be executed on more than one core. In GDDI, it is the number of cores per group that counts, so the user can parry an insolent un-HPC complaint of using too many cores by increasing the number of groups.

There are a few rarely used FMO runs that cannot employ GDDI fully, for example, the task of getting BDA corrections. In these cases, GDDI can be used with 1 group.

Another nasty feature that can be maddening is that there is a limit on the length of file names in GAMESS (different for each file), and this limit (on the order of 256 bytes) may be entirely hidden from the user, yet files whose names exceed the limit cannot be read. Although the allowance on file names is not very parsimonious, some users who love long names may be unpleasantly surprised at the result of their extravagant indulgence. GAMESS reads a few files in this way, including DFTB and force field parameters. A solution is to create a symbolic link if long names are much too dear to part with. A file name is usually absolute including all subdirectories.

6.2.4 *Scientific Failure*

A scientific failure occurs when the user is trying to use an inappropriate method. For example, trying to compute a system with a strong multireference character using a single-reference wave function. If there is a broken bond, and the user tries to calculate it with RHF, it often fails to converge, because the problem vociferously calls for a different method (UHF or MCSCF).

A common mistake is to use a poor molecular structure. In force fields, if the spring representing a bond is far stretched, it may go to its relaxed position when the geometry is optimized. But for QM methods, it is not so. If a structure has broken bonds, the calculation is not likely to converge (unless one attacks the problem with a proper multireference method). Not all QM methods are equal in their behavior regarding a poor structure. Some are more forgiving than others, and DFTB and HF-3c are among the most "omnivorous". If they work, one could optimize geometry with them first and refine with a better method later if desired.

Computing a protein in gas phase can have a very bad convergence due to its metallic character, but adding solvent tends to stabilize the electronic state. Not all scientific failures show themselves openly. Some poorly chosen methods do converge and produce some results, but they may be inappropriate or inaccurate.

It is the author's belief that when a calculation does not converge, it is the fault of the user (or bugs in the program). When either the geometry or the method is not appropriate, divergence can occur.

FMO has been designed to work with compact basis sets. Using it with diffuse functions is an example of an unduly stretch. It goes against the physics built in the commonly used FMO models. If one is to use diffuse functions in FMO, special efforts and techniques have to be used, as described in Section 5.3.

It is hard to give a simple advice on scientific failures with running GAMESS, other than stressing the need on the part of the user to educate oneself by lifelong learning, understanding the limitations of various methods, paying attention to choosing the right method for the problem on hand, starting with a good structure, and setting options properly.

6.2.5 *Input Errors*

To incorporate a comprehensive list of all input errors may require a multi-volume *Encyclopædia Ludorum*, so here only a small selection of common yet non-trivial errors are described. User errors are best dealt with when rerunning on 1 core to avoid losing important error messages.

Some input errors related to setting up the fragment boundaries may be resolved by doing a special run with NPRINT = 16 in $FMOPRP, which does a check of fragmentation. In defining INDAT, each atom in $FMOXYZ should be placed in exactly one fragment (segment), else an error will result.

In this Section, the problems of convergence are not dealt with, and the focus is on errors in input files, with the emphasis on FMO-specific problems. Some of the error messages are real head-scratchers, when even an experienced developer can gape in utter disbelief.

One relatively frequent problem is that for MODIO = 3072, MEM10 is set internally, which is a buffer for storing dictionary file for each *n*-mer. As it is not known what its size should be, 10 MW are allocated, which may not be enough when large fragments are used (in this case, an error message reports the memory shortage).

MEM10 can be explicitly set in words, for example, MEM10 = 20000000 (20 MW). With MODIO = 3072 and RUNTYP = OPTFMO, the user should set NFGD in $OPTFMO to the same value as NFRAG in $FMO.

In PCM, it is necessary to allocate memory for tesserae before the exact number is known. The guess is usually sufficient, but for radii sets in which hydrogen has a non-zero radius (such as VANDW), the guess may not suffice, in which case MXTS in $PCM should be set, based on the error messages. IDISP has a large scaling factor applied to atomic radii, generating more tesserae than for other terms.

Other errors are grouped in Table 95 (fragmentation) and Table 96 (the rest). For E1.1 (incomplete $DATA), calcium ($Z = 20$) has no basis set defined. Adding Ca to $DATA should resolve the problem. For E1.2 (incorrect INDAT), INDAT may be missing one atom, for example, $FMOXYZ has 20 atoms, but only 19 of them are used in INDAT. Some atom is not assigned to any fragment. E1.3 (basis set mismatch) occurs for systems with detached bonds, in which the basis set in $FMOHYB/$FMOBND does not match the actual basis set in $DATA (or in $BASIS). In this example, the number of AOs for a BDA in one basis is 15, and 5 in another (possibly, the user adjusted the basis set in $BASIS but failed to change $FMOBND and/or $FMOHYB).

E1.4 (invalid ESP for FMO1) occurs when MODGRD = 2 is set for FMO1, which is not allowed for gradients. For example, a non-FMO calculation is done as FMO with 1 fragment (as in OPTFMO). In this case, the suboption in MODGRD should not include 2 (for 1 fragment, MODGRD = 0 should be used).

Table 95.   Input errors related to fragmentation.

ID	Output sample
E1.1	Basis set not found for iz = 20, ibas = 1, iat = 1, ifg = 3, ilay = 1.
E1.2	Bad indat: nfg(indat,nfg) = 6 6 natfmo(indat,fmoxyz) = 19 20
E1.3	Basis set size mismatch in ROTCAO: 15 5
E1.4	ESP derivative (modgrd = 2) requires nbody > 1
E1.5	ESP derivative (modgrd = 2) requires modesp = 0
E1.6	Active domain is not separated from frozen.
E1.7	Atoms 2 1 may have to have a bond between them defined in $FMOBND.

E1.5 (invalid MODESP for ESP) occurs when a block-wise ESP (Section 5.2) is used for gradients, which is not allowed. A solution is to set MODESP = 0. E1.6 (FD domain error) occurs in FD, when there is a covalent bond between **A** and **F** domains, which is not allowed. A solution is to redefine domains. E1.7 (undeclared bonds between fragments) occurs if the user failed to declare a covalent bond between fragments; it can arise when a metal ion is so close to a solvent molecule that there appears to be a covalent bond between them, which is not defined in $FMOBND. A solution is to correct $FMOBND if appropriate or to set PRTDST(3) to a small number such as 0.4 to silence the whistleblower.

Table 96.   Other input errors.

ID	Output sample
E2.1	THE POINT GROUP OF THE MOLECULE IS C1
	THE ORDER OF THE PRINCIPAL AXIS IS 0
	**** ERROR READING VARIABLE IDUM CHECK COLUMN 2
	$END
E2.2	ERROR FOUND WHILE LOOKING FOR MAXANG FOR H-1
E2.3	$DFTBSK KEYWORD NOT FOUND
E2.4	GDDI ERROR: MORE GROUPS THAN NODES
E2.5	**** ERROR READING INPUT GROUP $FMO *****
	THE PROBLEM IS WITH THIS INPUT LINE, NEAR THE X MARKER

E2.1 in Table 96 (basis set undefined) is the champion in the user bewilderment. It occurs when a basis set is not defined anywhere (i.e., neither in $BASIS nor in $DATA). A solution is to define basis sets properly. The most charming part of this error message is its pointing a thundering finger at an $END, correct yet misleading. E2.2 (atom name mismatch) occurs when a layer index is used in $DATA for DFTB, which is not allowed. For example, "C-1" should be "C" in $DATA. E2.3 (PARAM missing) occurs when PARAM in $DFTB is not defined.

E2.4 (not enough nodes) is a common error in GDDI, which arises when the user asks for more groups than nodes. For example, in order to use 4 groups (NGRFMO(1) = 4 in $FMOPRP or NGROUP = 4 in $GDDI), GAMESS should be executed with at least 4 logical nodes. It is easy to deal with this problem: more logical nodes should be created by either adjusting the 5th argument to rungms (for running on 1 node with sockets, or in any MPI run) or passing a node file (sockets on multiple nodes) (see Sections 2.4.2 and 2.4.3). For MPI, the 5th argument to rungms can be used to define the logical node size (the simplest is to set it to 1, to have as many logical nodes as there are cores).

E2.5 (general input error) is a general error in reading an unformatted group that happens all the time, for example, when a keyword was put into a wrong group, or was misspelled (or spelled correctly when it should have been misspelled, as the mistakenly correct IFOLLOW instead of the correctly mistaken IFOLOW). Another reason may be specifying too many array elements, for example, if an array is defined to have 5 elements and the user tries to specify 7. One very nasty FMO-specific issue is overstretching the list format of INDAT, listing atoms one by one without using ranges, for example, INDAT(1) = 0, 1, 2, 3, 4, 5 instead of INDAT(1) = 0 1 −5, because INDAT has a limit to the list size. Not putting a separator in "1−5" instead of the correct "1 −5" is an error. The X marker does a good job in pointing out where the problem is.

6.2.6 *Version Change*

Any user with a longer than a short-time interest in GAMESS has to face the prospect of updating the program. The most likely outcome is that the same results will be obtained after an update, but four cases may occur in which results will differ.

(1) If there was a bug in the old version (see RELEASE.md) that was patched, then the new version is correct and all old results should be recalculated with the new version. (2) There may be a new feature, that is turned on by default, which was not there before. (3) The default options for an old feature may be changed. For (2) and (3), there may be a way to set up a parameter in the input so that the old results can be reproduced using the new version. A good way to find out is to execute a diff command on the old and new output files. This should reveal what input options may have changed, and the input can be modified to get back the old results with the new version. (4) A new bug may have appeared in the new version. The

old version may be correct and the new one not. Clearly, it is hard for a general user to deal with (1) and (4).

6.3 Collection of Sample Input and Output Files

Here, a small collection of complete input files with an enhanced output is provided. The molecular systems are deliberately chosen to be very small, *n*-hexane $CH_3-(CH_2)_4-CH_3$ or water tetramer $(H_2O)_4$.

The coordinates for *n*-hexane (Figure 27) are in Table 97 and for water tetramer in Table 98. The BDA/BAA pairs in the fragmentation of *n*-hexane were chosen to have a mirror symmetry (in agreement with the actual symmetry of the molecule), that is, "−2 3" and "−5 4", and not "−4 5" for the latter. This reduces the artefacts of fragmentation, although not by much.

All input files use 1 GDDI group, which is appropriate when executed on 1 CPU core. When more cores are used, the parallel options have to be amended accordingly. For a simple solution, all non-DFTB jobs can be executed as they are, even if multiple cores are used. DFTB inputs should be run as 1 core per group, which can be done by adding NSUBGR = −1 to $GDDI.

One important point about FMO output is that for connected fragments ($R_{IJ} = 0$), the value of the total interaction in the last column of the two-body FMO properties is printed as the quantity called an incomplete PIE.

$$\Delta \dot{E}_{IJ} = \begin{cases} \Delta E_{IJ}, & \text{for } R_{IJ} \neq 0 \\ \Delta E_{IJ} - \Delta E'_{IJ}, & \text{for } R_{IJ} = 0 \end{cases} \tag{93}$$

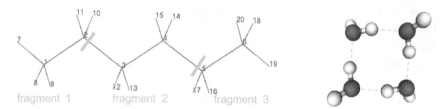

Figure 27. *n*-hexane and $(H_2O)_4$. Covalent boundaries are shown with green bars at BDAs.

Table 97.   Coordinates for *n*-hexane, optimized with DFT3/3ob/D3(BJ).

```
$FMOXYZ
C 6.0  -0.0090265397  -1.4528302351   1.3291779314
C 6.0  -0.0052718874  -0.0344038939   0.7770701004
C 6.0  -0.0021295722  -0.0019032344  -0.7538092935
C 6.0   0.0010196245   1.4245205078  -1.3075764541
C 6.0   0.0038158128   1.4558633140  -2.8384806613
C 6.0   0.0070079516   2.8736954139  -3.3921496155
H 1.0  -0.0114230777  -1.4472703483   2.4168154940
H 1.0   0.8694856469  -2.0019678667   0.9969284145
H 1.0  -0.8880522784  -1.9987581317   0.9930280448
H 1.0  -0.8809872730   0.5063913287   1.1505124346
H 1.0   0.8709765744   0.5030928848   1.1540529039
H 1.0   0.8732295905  -0.5425554874  -1.1288044016
H 1.0  -0.8776340229  -0.5399513123  -1.1322147646
H 1.0  -0.8742899254   1.9654709236  -0.9327957863
H 1.0   0.8766376336   1.9627980526  -0.9296640543
H 1.0  -0.8723439347   0.9175913153  -3.2145018401
H 1.0   0.8796956515   0.9149516428  -3.2113684699
H 1.0  -0.8715793183   3.4229749291  -3.0603195183
H 1.0   0.0088734121   2.8669605844  -4.4798039114
H 1.0   0.8859959318   3.4203296126  -3.0570965527
$END
```

Table 98.   Coordinates for water tetramer, optimized with DFT3/3ob/D3(BJ).

```
$FMOXYZ
O 8.0  -0.1126896491  -0.2165469386  -0.1035698838
H 1.0  -0.3864165764   0.3338585097  -0.8419424742
H 1.0   0.8428111505  -0.0780118199   0.0761444269
O 8.0   2.5837169828  -0.2098127302   0.2118445053
H 1.0   3.1388399861   0.1064680960   0.9294371017
H 1.0   2.7217579562  -1.1742214748   0.0869720190
O 8.0   2.5929115943  -2.9186704543   0.0417264762
H 1.0   1.6284247796  -3.0492699736   0.1728309330
H 1.0   2.9078951856  -3.5053393525  -0.6508676312
O 8.0  -0.1168848748  -2.9135794830   0.2028922197
H 1.0  -0.2461973358  -1.9570082413   0.0214612021
H 1.0  -0.7115311992  -3.1937861376   0.9035061053
$END
```

This complication is needed because ΔE_{IJ} and $\Delta E'_{IJ}$ are huge (~9000 kcal/mol) for connected dimers. However, in PIEDA all PIEs are always printed as ΔE_{IJ}. To obtain ΔE_{IJ} for connected dimers in a non-PIEDA run, the printed contributions can be summed according to Eq. (4), or the value can be obtained as $\Delta E_{IJ} = \Delta \dot{E}_{IJ} + \Delta E'_{IJ}$, with the same result.

6.3.1 *FMO2-RHF/STO-3G, HOP*

The input header is given in Table 99 and the coordinates should be added from Table 97. The fragmentation here follows the list format,

Table 99. *n*-hexane, FMO2-RHF/STO-3G, HOP, ENERGY.

```
$CONTRL NPRINT = -5 RUNTYP = ENERGY SCFTYP = RHF $END
$SYSTEM MWORDS = 100 MEMDDI = 100 $END
$GDDI NGROUP = 1 $END
$SCF DIRSCF = .T. NPUNCH = 0 $END
$BASIS GBASIS = STO NGAUSS = 3 $END
$FMO NBODY = 2 NLAYER = 1 NFRAG = 3
INDAT(1) =  0 1 -2 7 -9 10 -11 0 3 -4 12 -13 14 -15 0 5 -6 16 -17 18 -20 0
$END
$FMOPRP NPRINT = 9 $END
$FMOBND
-2 3 STO-3G
-5 4 STO-3G
$END
$FMOHYB
STO-3G 5 5
1 0 -0.117784 0.542251 0.000000 0.000000 0.850774
0 1 -0.117787 0.542269 0.802107 0.000000 -0.283586
0 1 -0.117787 0.542269 -0.401054 -0.694646 -0.283586
0 1 -0.117787 0.542269 -0.401054 0.694646 -0.283586
0 1 1.003621 -0.015003 0.000000 0.000000 0.000000
$END
$DATA
DFTB3/3OB E = -15.5992436830 GMAX = 0.0000744 GRMS = 0.0000230
C1
H-1 1
C-1 6
$END
```

with the first element of INDAT set to 0. Atoms 1–2, 7–9, and 10–11 are in fragment one, that is, CH_3–CH_2– (see Figure 27). The basis set is built-in STO-3G (to match it, HMOs for STO-3G are placed in $FMOBND).

NPRINT = –5 in $CONTRL and NPUNCH = 0 in $SCF reduce the output and DAT files, respectively. The packed option NPRINT = 9 in $FMOPRP defines the output level 1 with Mulliken charges (8) printed (8 + 1 = 9).

The comment in $DATA (the first line) shows how the geometry was obtained, which is a good style (in this case it shows the level of geometry optimization, with a reference energy and gradient).

The results are shown in Figure 28. Here and in other similar Figures, some truncation and visualization enhancement of output was done, including a truncation of trailing significant figures.

```
One-body FMO properties.

    I                 E'_I (a.u.)    d_x^I      d_y^I      d_z^I  (Debye)

1(frg00001,L1) -63.341351385  0.00496  0.08131 -2.37858
2(frg00002,L1) -76.282892568 -0.00052 -0.00181 -0.00011
3(frg00003,L1) -63.341340283 -0.00441 -0.07931  2.37882
  Total energy of the molecule: Euncorr(1)= E^FMO1= -202.965584237
  Dipole moment D(xyz),DA(1)= d^FMO1= 0.00003 0.00018 0.00013 0.00023ª
Two-body FMO properties.

I J DL Z  R_IJ    ΔQ_I→J       E'_IJ         ΔE'_IJ     ΔE_IJ^V   ΔE_IJ^solv  ΔĖ_IJ

2 1 N1 0 0.00   0.0115 -154.4604273 -14.836183 0.004013 0.000 2.518
3 1 N1 0 1.30  -0.0000 -126.6785502   0.004141 0.000012 0.000 2.606
3 2 N1 0 0.00  -0.0115 -154.4605011 -14.836268 0.004018 0.000 2.522
  Total energy of the molecule: Euncorr(2)= E^FMO2= -232.625850133
  Dipole moment D(xyz),DA(2)= d^FMO2= 0.00000 -0.00013 -0.00007 0.00015ª
n-body Mulliken  atomic charges Q(n)
   A   I   Z       Q_A^FMO1      Q_A^FMO2
   1   1  6.0    -0.182189    -0.177380
   2   1  6.0     0.108048    -0.089705
...
```
ª Listed as d_x, d_y, d_z, and $|d|$ (Debye).

Figure 28. Enhanced (truncated and highlighted) output for the FMO2-RHF input in Table 99.

The purpose of enhancement is to show how to read output and find quantities of interest, aided by mathematical symbols.

As explained above, for connected dimers ($\Delta R_{IJ} = 0$), $\Delta \dot{E}_{IJ}$ is printed in the last column of two-body properties (in kcal/mol) instead of ΔE_{IJ}. For example, for dimer 2,1, $\Delta \dot{E}_{IJ} = 2.518$ kcal/mol is equal to $\Delta E_{IJ}^{V} = 0.004013$ hartree, whereas $\Delta E_{IJ}^{r} = -14.836183$ hartree.

The units of energies are not always specified. Many are in hartree, but some are in kcal/mol. As a general convention, energies printed with 3 or fewer significant figures after the decimal point are in kcal/mol and with 8 or more, in hartree.

6.3.2 *FMO2-DFT/3-21G, AFO, LCMOX*

The input header is given in Table 100 and the coordinates should be added from Table 97 (the last atom #20 is to be removed).

Table 100. *n*-hexyl cation, FMO2-DFT/3-21G, AFO, LCMOX.[a]

```
$CONTRL NPRINT = -5 RUNTYP = ENERGY SCFTYP = RHF
DFTTYP = B3LYP LOCAL = RUEDNBRG MAXIT = 90 $END
$SYSTEM MWORDS = 100 MEMDDI = 100 $END
$GDDI NGROUP = 1 $END
$SCF DIRSCF = .T. NPUNCH = 0 $END
$BASIS GBASIS = N21 NGAUSS = 3 $END
$FMO NBODY = 2 NLAYER = 1 NFRAG = 3 ICHARG(1) = 0,0,1 RAFO(1) = 1,1,1
INDAT(1) =  1 1 2 2 3 3 1 1 1 1 1 2 2 2 2 3 3 3 3 $END
$FMOPRP NPRINT = 9 NLCMO(1) = 15,0,0 $END
$FMOBND
-2 3
-5 4 NONE 1
$END
$DATA
DFTB3/3OB E = -15.5992436830 GMAX = 0.0000744 GRMS = 0.0000230
C1
H-1 1
C-1 6
$END
```

[a]The last hydrogen atom #20 in *n*-hexane was removed, producing a cation.

The combination RHF + B3LYP means R-B3LYP (spin restricted RDFT). As DFT often has a bad convergence, the maximum number of iterations is increased to 90.

The fragmentation is given in the index format in INDAT, the same fragments as in Table 99, but in a different style of INDAT. The charges of each fragment are set in ICHARG, and the last fragment is a cation (in Table 99 the charges are not defined, so their default values of 0 are used).

AFO is chosen by specifying RAFO and a localization in LOCAL (here RUEDNBRG). The large model systems are used (the RAFO values are 1,1,1), $-CH_2-CH_2-$ plus caps on both sides. The model system for the second detached bond includes a cationic terminus $-CH_2^+$. Its charge (+1) should be given in $FMOBND. The format of this group requires a name between the BDA/BAA indices (5 4) and the charge. The name is ignored but should be given, so that here "–5 4 NONE 1" is used (the last number is the charge +1). The charges of model systems by default are 0, so the charge for the first model system of the "–2 3" BDA/BAA is not specified. The user may not know the charge of the model system. An easy approach is to set no charge (all neutral) and run. If the charge is mistaken by an odd number (by 1, 3, etc.), there will be an error message; for mistakes of an even value, no error is detected. However, a mistake of +1 or –1 is the most common. The coordinates of each model system are printed and can be inspected when in doubt.

For FMO2, LCMOX is more accurate than LCMO, chosen with NLCMO(1) = 15,0,0 (for FMO3, LCMO should be used with 13,0,0). In the first option of 15 = 1 + 2 + 4 + 8, 1 turns on LCMO, 2 adds X, 4 is a technical option, and 8 prints MOs. The Fock and overlap matrices may be useful for computing the rates of charge transport between fragments in the Marcus theory. They can be printed by adding 16 (Fock) and 32 (overlap) to NLCMO(1). The other two integers, 0,0, mean that all occupied and virtual orbitals in each monomer are used. The results are shown in Figure 29.

In LCMO, there is an internal basis set redundancy problem due to virtual orbitals for detached bonds, so warnings are printed about small overlaps. After a diagonalization, the eigenvalues and,

```
LCMO diagonalisation requires       8704 words.
..... WARNING .....
THE SMALLEST EIGENVALUE OF THE OVERLAP MATRIX IS     -0.00000000
THERE IS(ARE)    8 EIGENVALUE(S) LESS THAN   1.00E-06
THE NUMBER OF CANONICAL ORBITALS KEPT IS     80
LCMO linear dimensions:      88    80 full     80
Monomer MO and AO dimensions matched!
          -----------
          EIGENVALUES     = MO energies (hartree)
          -----------
    1 A       2 A       3 A       4 A       5 A       6 A
-10.483712 -10.367281 -10.302178 -10.259607 -10.233593 -10.207318
...

          -------------
          EIGENVECTORS    = Molecular orbitals
          -------------
                    1         2         3         4         5
                -10.4837   -10.3673  -10.3022  -10.2596  -10.2336
                    A         A         A         A         A
1  C  1  S     0.000256  -0.003806 -0.000277 -0.003574  0.015493
2  C  1  S    -0.000164  -0.000171 -0.001199 -0.001512  0.004530
...

Two-body FMO properties.
```

I	J	DL	Z	R_{IJ}	$\Delta Q_{I \rightarrow J}$	E'_{IJ}	$\Delta E'_{IJ}$	ΔE^V_{IJ}	ΔE^{solv}_{IJ}	$\Delta \dot{E}_{IJ}$
2	1	N1	0	0.00	0.1534	-156.6060209	-14.633265	-0.009855	0.000	-6.184
3	1	N1	0	1.30	0.0493	-127.4692144	0.005397	-0.000126	0.000	3.308
3	2	N1	0	0.00	-0.0189	-155.6993880	-14.752064	-0.007539	0.000	-4.731

```
 Total energy of the molecule: Euncorr(2)= E^FMO2 =  -234.594799308
Dipole moment D(xyz),DA(2)= d^FMO2 = -0.56969   6.02248 -8.00045 10.03005
```

Figure 29. Enhanced output for the FMO2-DFT/LCMO input in Table 100.

optionally, eigenvectors are printed. Labels A refer to the totally symmetric irreducible representation in C_1 point group. Orbital energies are in hartree.

6.3.3 *FMO2-UHF/6-31G*/D3(BJ)*

The input header is given in Table 101 and the coordinates should be added from Table 98. D3(BJ) is chosen by adding DC = .T. IDCVER = 4 (these options in $DFT also apply to HF).

Table 101.   Water tetramer, FMO2-UHF/6-31G*/D3(BJ), GRADIENT.[a]

```
$CONTRL NPRINT = -5 RUNTYP = GRADIENT SCFTYP = UHF
MAXIT = 90 ISPHER = 1 $END
$SYSTEM MWORDS = 100 MEMDDI = 100 $END
$GDDI NGROUP = 1 $END
$SCF DIRSCF = .T. NPUNCH = 0 $END
$BASIS GBASIS = N31 NGAUSS = 6 NDFUNC = 1 $END
$DFT DC = .T. IDCVER = 4 $END
$FMO NBODY = 2 NLAYER = 1 NFRAG = 4 NACUT = 3 MODGRD = 42
SCFFRG(1) = RHF,RHF,RHF,UHF MULT(1) = 1,1,1,1 $END
$FMOPRP NPRINT = 8 $END
$GUESS MIX = .T. $END
$DATA
DFTB3/3OB E = -16.3313609921 GMAX = 0.0000625 GRMS = 0.0000231
C1
H-1 1
O-1 8
$END
```

[a]One water is a UHF singlet, others are RHF singlets.

Here, the automatic fragmentation is used (NACUT = 3 atoms per fragment), so INDAT is not needed. MODGRD = 42 is set to obtain accurate gradients (available for HF/HOP). The number of iterations is increased to 90 (MAXIT) because UHF often has a poor convergence.

ISPHER = 1 is set to use pure spherical AOs, which is recommended for any basis set with $l = 2$ (d) or higher angular momentum (in 6-31G*, carbon has a d-function). There is a negligible extra cost in setting ISPHER = 1 for pure s,p basis sets (which are rare anyway), and the user can just set ISPHER = 1 for any basis set.

For runs with empirical dispersion (D), the following quantities are printed (exemplified here for UHF).

$$E^{FMOi-UHF-D} = E^{FMOi-UHF} + E^{FMOi-D} \qquad (94)$$

where the UHF-D energy is split into UHF and D energies.

Pair interactions are split as

$$\Delta E_{IJ} = \Delta E_{IJ}'^{UHF-D} + \Delta E_{IJ}^{V} + \Delta E_{IJ}^{solv} \qquad (95)$$

where the internal contributions to PIEs are

$$\Delta E_{IJ}^{\prime\text{UHF-D}} = \Delta E_{IJ}^{\prime\text{UHF}} + \Delta E_{IJ}^{\text{DI}} \tag{96}$$

```
One-body FMO properties.
```

I	$E_I^{\prime\text{UHF-D}}$ (a.u.)	$E_I^{\prime\text{UHF}}$ (a.u.)	d_x^I	d_y^I	d_z^I (Debye)
1(frg00001,L1)	−76.0083358	−76.0038635	1.61014	1.57522	−1.08683
2(frg00002,L1)	−76.0083272	−76.0038549	1.58309	−1.54457	1.16933

```
...
```

```
    Total energy of the molecule: Eunco+D(1)= E^FMO1-UHF-D = -304.033321875

    Total energy of the molecule: Euncorr(1)= E^FMO1-UHF =     -304.015432713

    Total energy of the molecule: Edisp  (1)= E^FMO1-D =       -0.017889162

    Dipole moment D(xyz),DA(1)= d^FMO1 = 0.00256 -0.00016 -0.01566 0.01587
Two-body FMO properties.
```

I	J	DL	Z	R_{IJ}	$\Delta Q_{I\to J}$	$E_{IJ}^{\prime\text{UHF-D}}$	$E_{IJ}^{\prime\text{UHF}}$	$\Delta E_{IJ}^{\prime\text{UHF-D}}$	$\Delta E_{IJ}^{\prime\text{UHF}}$	ΔE_{IJ}^{V}	$\Delta E_{IJ}^{\text{solv}}$	ΔE_{IJ}
2	1	N1	0	0.67	−0.03	−152.032	−152.019	−0.015	−0.011	−0.0020	0.000	−10.9
3	1	N1	0	1.28	0.00	−152.021	−152.012	−0.004	−0.004	−0.0005	0.000	−3.4

```
...
```

```
    Total energy of the molecule: Eunco+D(2)= E^FMO2-UHF-D = -304.113972067

    Total energy of the molecule: Euncorr(2)= E^FMO2-UHF =   -304.080857736

    Total energy of the molecule: Edisp  (2)= E^FMO2-D =     -0.033114331

    Dipole moment D(xyz),DA(2)= d^FMO2 = 0.00209 -0.00035 -0.01586 0.01601
Energy gradient (hartree/bohr), no BSSE: G(2)
```

A	I	Z	$dE^{\text{FMO2-UHF-D}}/dR_{Ax}$	$dE^{\text{FMO2-UHF-D}}/dR_{Ay}$	$dE^{\text{FMO2-UHF-D}}/dR_{Az}$ (hartree/bohr)
1	1	8.0	−0.008960274	−0.007522903	0.002604128
2	1	1.0	−0.010316316	0.008522819	−0.011659913

```
...
 (2) MAXIMUM GRADIENT = 0.0194036    RMS GRADIENT = 0.0099108
```

Figure 30.   Enhanced output for the FMO2-UHF input in Table 101.

SCFTYP in $CONTRL should be set to a non-RHF type, if any is used, so SCFTYP is set to UHF. Out of 4 fragments, only the last one is treated with UHF, and the rest with RHF (SCFFRG). The multiplicity of each fragment is 1, that is, all are singlets. A UHF singlet requires mixing initial α and β spin orbitals, so MIX = .T. is used. The most verbose output level 0 is required to print spin properties in a gradient run so that combining 0 with 8 for printing Mulliken charges, NPRINT = 8 in $FMOPRP, is set. The results are shown in Figure 30.

6.3.4 *FMO3-CCSD(T)/AP/cc-pVDZ:aug-cc-pVDZ*

The input header is given in Table 102 and the coordinates should be added from Table 98. Because diffuse functions tend to have a slow

Table 102. Water tetramer, FMO3-CCSD(T)/AP/cc-pvdz:aug-cc-pvdz, ENERGY.

```
$CONTRL NPRINT = −5 RUNTYP = ENERGY CCTYP = CCSD(T) MAXIT = 90
ISPHER = 1 QMTTOL = 1E-6 $END
$SYSTEM MWORDS = 200 MEMDDI = 100 $END
$GDDI NGROUP = 1 $END
$SCF DIRSCF = .T. NPUNCH = 0 $END
$FMO NBODY = 3 NLAYER = 1 NFRAG = 4 NACUT = 3 NDUALB = 1 $END
$FMOPRP NPRINT = 9 $END
$DATA
DFTB3/3OB E = −16.3313609921 GMAX = 0.0000625 GRMS = 0.0000231
C1
H.1 1
CCD

O.1 8
CCD

H.2 1
ACCD

O.2 8
ACCD

$END
```

```
Total properties for the dual basis FMO.
```
Total energy of the molecule: Ep1   (1)= $\Delta E^{\text{PL1,CCSD(T)}}$ = 0.018705076

Total energy of the molecule: Ep1dunc(1)= $\Delta E^{\text{PL1,RHF}}$ =   0.013430950

I J $\Delta E_{IJ}^{\text{AP}}$ = $\Delta E_{IJ}^{\text{BS2,0}} + \Delta E_{IJ}^{\text{PL}}$ = $\Delta E_{IJ}^{\text{BS1,}V} + \Delta E_{IJ}^{\text{BS}}$ (kcal/mol)

```
2 1  -9.559  -4.777 -4.782  -11.453 1.895
3 1  -3.252  -1.817 -1.435   -3.697 0.445
...
```

Total energy of the molecule: Ep1   (2)= $\Delta E^{\text{PL2,CCSD(T)}}$ = -0.016405020

Total energy of the molecule: Ep1dunc(2)= $\Delta E^{\text{PL2,RHF}}$ =   -0.014870649

Total energy of the molecule: Ep1   (3)= $\Delta E^{\text{PL3,CCSD(T)}}$ = -0.001209937

Total energy of the molecule: Ep1dunc(3)= $\Delta E^{\text{PL3,RHF}}$ =   -0.000949256

Total energy of the molecule: Ecorr (3)= $\Delta E^{\text{FMO3-CCSD(T)/AP}}$ = -305.13844983[a]

Total energy of the molecule: Euncorr(3)= $\Delta E^{\text{FMO3-RHF/AP}}$ = -304.19222830

Total energy of the molecule: Edelta (3)= $\Delta E^{\text{FMO3-CCSD(T)/AP}} - \Delta E^{\text{FMO3-RHF/AP}}$ =...

Dipole moment D(xyz),DA(3)= $\mathbf{d}^{\text{FMO3/AP}}$ = 0.0019 -0.00041 -0.0147 0.0148

```
Atomic charges Q
```
A I Z $Q_A^{\text{FMO3-RHF/AP}}$ [b]
```
1 1 8.0   -0.503489
2 1 1.0    0.169572
...
```
---
[a] This is the total FMO/AP energy at the CCSD(T) level.
[b] There are no CC charges (instead, HF charges are computed).

Figure 31.   Enhanced output for the FMO3-CCSD(T)/AP input in Table 102.

converge, MAXIT was increased to 90. ISPHER = 1 is required for Dunning basis sets (cc-pnDZ and aug-cc-pnDZ). QMTTOL is set to 10^{-6} to eliminate possible linear dependencies, which are common for rings (absent in this system, but set for a general case).

AP is selected with NDUALB = 1, and the two basis sets for it are set in $DATA for each atom. FMO3 is chosen with NBODY = 3. The results are shown in Figure 31. For MWORDS = 100 two integral passes are required (as can be learned from a run with MWORDS = 100 by reading the output), so the memory amount is increased to 200.

$\Delta E^{\text{PL}m}$ is defined in Eq. (86). Here, these values are computed for $m \le n$ in FMOn, at the RHF and CC levels. The total AP values (combining 3 calculations) are denoted by AP in the superscript.

6.3.5 *FMO2-MCSCF/6-31G*

For MCSCF, a preliminary RHF is usually done. The purpose of it is to prepare a set of MOs for an active space. For example, for a transition metal with partially occupied s and d orbitals, one can pick 6 orbitals, and for an organic system, pairs of bonding and antibonding orbitals. Here, as an example, n-hexene is computed, by including π- and π*-like orbitals (here, because the geometry of –CH=CH$_2$ is not planar, the system has a diradical character).

The first step is FMO-RHF. It is enough to use FMO1, obtain orbitals of one fragment (#3), and select an active space for this MCSCF fragment. Other fragments are treated with RHF (and no orbitals should be prepared for them). The input for FMO-RHF is not shown (see Table 99 for a similar input); however, two points are to be noted. NPRINT = –5 is removed from $CONTRL so that the output includes MOs, which may be inspected (see Section 4.8 for plotting them), and NPUNCH = 0 is removed from $SCF so that MOs are written to the DAT file (that is important).

For this system, the choice was made to take HOMO and LUMO as the active orbitals. In this case, no orbital reordering is needed, because all MOs are already in the right order of core, active, and virtual. Otherwise, one would have to change the order of the orbitals (manually, see Section 5.5.3). To make an MCSCF input file, the input from RHF is used as a base with some changes.

The $VEC group for monomer #3 from the DAT file of FMO-RHF should be taken (open the DAT file in a text editor and look for "FMO ORBITALS   3   0   0"). Due to multiple monomer iterations, there may be many $VEC groups for fragment #3. In this case, the $VEC from the last monomer iteration is the best. If multiple groups are used in GDDI, then the user should identify the group that did the last monomer calculation by checking all outputs, and extract MOs from the DAT file of that group. It is possible to print only the final set of MOs (accomplished with NPRINT = –5 in $CONTRL and NOPFRG(3) = 1 in $FMO). These RHF orbitals of fragment #3 in $VEC are copied to the MCSCF input, and the group is renamed to $VEC1 (it is possible to read many $VEC groups in one FMO calculation, for example, also for dimers in case of poor convergence).

Table 103.   *n*-hexene, FMO2-MCSCF/6-31G.[a]

```
$CONTRL NPRINT = -5 RUNTYP = ENERGY SCFTYP = MCSCF $END
$SYSTEM MWORDS = 100 MEMDDI = 100 $END
$GDDI NGROUP = 1 $END
$SCF DIRSCF = .T. NPUNCH = 0 $END
$MCSCF FULLNR = .T. SOSCF = .F. CANONC = .F. $END
$DET NCORE = 6 NACT = 2 NELS = 2 $END
$BASIS GBASIS = N31 NGAUSS = 6 $END
$FMO NBODY = 2 NLAYER = 1 NFRAG = 3 SCFFRG(1) = RHF,RHF,MCSCF
INDAT(1) =  1 1 2 2 3 3 1 1 1 1 1 2 2 2 2 3 3 3 $END
$FMOPRP NPRINT = 9 IJVEC(1) = 3,0,0,1,24 NGUESS = 18 $END
$FMOBND
-2 3 6-31G
-5 4 6-31G
$END
$FMOHYB
6-31G 9 5
...
$END
$DATA
DFTB3/3OB E = -15.5992436830 GMAX = 0.0000744 GRMS = 0.0000230
C1
H-1 1
C-1 6
$END
$VEC1
 1  1 8.81103621E-02-1.45217156E-04-1.03828893E-04-2.44604555E-05-
3.03760392E-03
 1  2 1.29898754E-03 5.56951606E-05 1.98209407E-03-2.46416842E-03
9.92440173E-01
...
$END
```

[a]Two hydrogen atoms (#17 and #20) were removed from *n*-hexane, changing -CH$_2$-CH$_3$ into a twisted -CH = CH$_2$ in *n*-hexene.

The MCSCF input is shown in Table 103 and the coordinates are in Table 97 (two H atoms, 17 and 20, should be removed). In any FMO-MCSCF, there are both RHF and MCSCF fragments, therefore, both $SCF and $MCSCF are relevant. A CI basis is not set in CISTEP of $MCSCF, so that the default (ALDET) is used, for which $DET should define the active space.

There are NCORE = 6 core and NACT = 2 active orbitals with NELS = 2 electrons in the active space. This is the minimum active space possible, corresponding to correlating just 2 electrons in 2 orbitals, which is equivalent to generalized valence bond (GVB) theory.

The wave functions for fragments are specified in SCFFRG, two RHF fragments and one MCSCF. IJVEC(1) = 3,0,0,1,24 is used to read orbitals, where 3,0,0 define an *n*-mer *I,J,K*. Ignoring zeroes, *I* = 3, so $VEC1 is used for fragment number 3. The fourth index is the layer number, 1. The last value (24) is the number of MOs in $VEC1. For MCSCF all MOs should be read, including all virtuals, whereas for RHF it is enough to provide occupied orbitals.

For thus constructed input, MCSCF dimers do not converge (attributed to an interference of close detached bonds). A solution to this, especially for MCSCF fragments with covalent boundaries (as here), is to add NGUESS = 18 in $FMOPRP, which is a packed option, split as 2 and 16. 2 is the default value for NGUESS, meaning that dimer/trimer initial orbitals are automatically constructed as a union of monomer properties (the alternative is to use Hückel guess if 2 is not added).

16 is the option to do a preliminary RHF for dimers/trimers. In this case, just starting off from the union of monomer MOs leads to divergence. It is often helpful to do an RHF precalculation for each MCSCF dimer first, followed by an automatic overlap-based matching scheme to find the MOs among RHF dimer orbitals that resemble monomer MCSCF orbitals. Alternatively, one can try to play with convergence options in $MCSCF or read in dimer orbitals as $VEC*i*.

The results are shown in Figure 32. There are 4 determinants with substantial coefficients (in this case, the total number is also 4). Determinants are printed as their MO occupations, 1 or 0, separately for α and β spin-orbitals. The CI coefficients show that the second active orbital has a substantial population, which is because this is a diradical rather than a π-bond system, as determined by the non-planar geometry.

The DL column shows the type of the dimer (D) and the layer L. For MCSCF dimers, D = M (multiconfigurational), and for RHF dimers, D = N (short for non-correlated). Z is the product of fragment charges of the two monomers.

```
CI EIGENVECTORS WILL BE LABELED IN GROUP=C1
PRINTING ALL NON-ZERO CI COEFFICIENTS


STATE 1 ENERGY=-73.8776808365  S= 0.00  SZ= 0.00  SPACE SYM=A ᵃ
 ALP| BET| CI COEFFICIENT
----|----|---------------
πα π*α| πβ π*β
 10 |10  | 0.8609183688  closed-shell RHF-like determinant for π²
 01 |01  |-0.5085173421  corresponds to (π*)²
 10 |01  |-0.0107162299
 01 |10  |-0.0107162299
...
```

Two-body FMO properties.

I	J	DL	Z	R_{IJ}	$\Delta Q_{I \rightarrow J}$	E'_{IJ}	$\Delta E'_{IJ}$	ΔE^V_{IJ}	ΔE^{solv}_{IJ}	$\Delta \dot{E}_{IJ}$ [b]
2	1	N1	0	0.00	0.0337	-156.2107443	-15.185758	0.001421	0.000	0.892
3	1	M1	0	1.30	-0.0000	-126.4669700	0.004755	-0.001468	0.000	2.062
3	2	M1	0	0.00	-0.0299	-154.9609885	-15.181121	0.000768	0.000	0.482

Total energy of the molecule: Euncorr(2)= $E^{\text{FMO2-MCSCF}}$ = -233.999693340

---

[a] The wave function in terms of determinants is printed for each n-mer separately (no total summary). Here, the monomer results are shown.

[b] Following the convention to exclude $\Delta E'_{IJ}$ from the last column for connected dimers, so that ΔE^V_{IJ} =0.001421 hartree = $\Delta \dot{E}_{IJ}$ =0.892 kcal/mol for dimer 21 (in vacuum, ΔE^{solv}_{IJ} =0).

Figure 32.   Enhanced output for the FMO2-MCSCF input in Table 103.

6.3.6 *FMO2-DFTB3/SMD, PA*

The input header is given in Table 104, the coordinates should be added from Table 98 and the PDB data should be taken from Table 105. Most analyses such as PA are conducted for RUNTYP = ENERGY. DFTB is selected by specifying a DFTB basis set in $BASIS. DFTB3 is chosen by selecting SCC = .T. DFTB3 = .T. DAMPXH = .T. in $DFTB. The parameter set is defined as PARAM. To match it, HMOs are chosen in $FMOBND and $FMOHYB. The D3(BJ) dispersion is selected in $DFT with DC = .T. IDCVER = 4. Note that atomic names in $DATA have no layer number, i.e., "C", not "C-1", as required for DFTB.

Table 104. *n*-hexane, FMO2-DFTB3/D3(BJ)/SMD, PA.[a]

```
$CONTRL NPRINT = –5 RUNTYP = ENERGY $END
$SYSTEM MWORDS = 100 MEMDDI = 100 $END
$GDDI NGROUP = 1 $END
$SCF DIRSCF = .T. NPUNCH = 0 $END
$BASIS GBASIS = DFTB $END
$DFTB SCC = .T. DAMPXH = .T. DFTB3 = .TRUE. PARAM = 3OB-3-1 $END
$DFT DC = .T. IDCVER = 4 $END
$PCM SOLVNT = WATER IEF = -10 SMD = .T. IFMO = -1 MODPAR = 73 $END
$FMO NBODY = 2 NLAYER = 1 NFRAG = 3
INDAT(1) =  0 1 –2 7 –11 0 3 –4 12 –15 0 5 –6 16 –20 0
INDATP(1) = –6
$END
INDATP(1) = 0 1 7 –9 0 2 10 11 0 3 12 13 0 4 14 15 0 5 16 17 0 6 18 –20 0
$FMOPRP NPRINT = 9 MODPAR = 8205 MODPAN = 1 $END
$FMOBND
–2 3 3OB-3-1
–5 4 3OB-3-1
$END
$FMOHYB
3OB-3-1 4 4
1 0 0.565124    0.000000    0.000000   –0.825006
0 1 0.565123    0.777823    0.000000    0.275002
0 1 0.565124   –0.388911    0.673614    0.275003
0 1 0.565124   –0.388911   –0.673614    0.275003
$END
$DATA
DFTB3/3OB E = –15.5992436830 GMAX = 0.0000744 GRMS = 0.0000230
C1
H 1
C 6
$END
```

[a]The PDB group in Table 105 should be added to this input. The blue line is a comment because it is outside of any group.

SMD is selected with SMD = .T. in $PCM. Note the absence of options incompatible with SMD such as ICOMP, ICAV, and IDISP (radii are not explicitly set, as required). The solvent is water (SOLVNT, truncated to the limit of 6 characters). IEF = –10 selects the iterative C-PCM solver. MODPAR = 73 in $PCM defines the new (partial) screening model, the only one fully supported in PA (73 =

Table 105.   Example of a $PDB group used in PA in Table 104.

$PDB										
HETATM	1	C	CH3	1	−0.009	−1.453	1.329	0	0	C
HETATM	2	C	CH2	2	−0.005	−0.034	0.777	0	0	C
HETATM	3	C	CH2	3	−0.002	−0.002	−0.754	0	0	C
HETATM	4	C	CH2	4	0.001	1.425	−1.308	0	0	C
HETATM	5	C	CH2	5	0.004	1.456	−2.838	0	0	C
HETATM	6	C	CH3	6	0.007	2.874	−3.392	0	0	C
HETATM	7	H	CH3	1	−0.011	−1.447	2.417	0	0	H
HETATM	8	H	CH3	1	0.869	−2.002	0.997	0	0	H
HETATM	9	H	CH3	1	−0.888	−1.999	0.993	0	0	H
HETATM	10	H	CH2	2	−0.881	0.506	1.151	0	0	H
HETATM	11	H	CH2	2	0.871	0.503	1.154	0	0	H
HETATM	12	H	CH2	3	0.873	−0.543	−1.129	0	0	H
HETATM	13	H	CH2	3	−0.878	−0.540	−1.132	0	0	H
HETATM	14	H	CH2	4	−0.874	1.965	−0.933	0	0	H
HETATM	15	H	CH2	4	0.877	1.963	−0.930	0	0	H
HETATM	16	H	CH2	5	−0.872	0.918	−3.215	0	0	H
HETATM	17	H	CH2	5	0.880	0.915	−3.211	0	0	H
HETATM	18	H	CH3	6	−0.872	3.423	−3.060	0	0	H
HETATM	19	H	CH3	6	0.009	2.867	−4.480	0	0	H
HETATM	20	H	CH3	6	0.886	3.420	−3.057	0	0	H
$END										

1 + 8 + 64, where 8 is the new model and 64 is a general PCM acceleration). IFMO = −1 chooses PCM<1>. $FMOPRP MODPAR = 8205 breaks down to 1 + 4 + 8 + 8192, of which 13 is the default option and 8192 is added to accelerate the screening model.

PA is chosen with MODPAN = 1. Because 32 is not added to it, segments are to be defined explicitly (with 32 they would have been identical to fragments). PA is typically used with NPRINT = 9. The number and the composition of segments are completely independent of the number and composition of fragments.

INDATP(1) = −6 means that segments are defined in $PDB (any negative number means that). A negative value passed in INDATP(1) is only used for memory allocation; its absolute value (6) may be equal or larger to the actual number of segments, which is determined by the $PDB; it is normal to specify a large value such as INDATP(1) = −1000 if the user is unsure of the number segments, which may happen for splitting side chains. In the input file here, the user pledges to have 6 or fewer segments in $PDB by specifying INDATP(1) = −6.

An alternative way of the same segment definition is shown as INDATP(1) in the list format of INDAT. Because this INDATP is outside of any input group, it is treated as a comment. For an alternative INDATP, with a non-negative first number, the number of segments is determined from the list of indices in the option.

Segments are defined inside the $PDB group, which uses the PDB format (only ATOM and HETATM entries are processed, others are ignored). Three columns are used in $PDB: the atom name (used only as a label), the segment label, and segment ID (the most important data). Here, atom names are C and H, and segment labels are CH_3 and CH_2.

The most important part of $PDB is the segment ID column (the fifth column in $PDB). It defines the segment to which each atom is assigned to. In this case, the column has numbers from 1 to 6, so that there are 6 segments. These values may start from any positive number, not necessarily from 1, for instance, if the PDB describes a truncated part of a protein, and the user likes to keep the original numbering. Internally, segments are renumbered starting from 1. Segment names are made by concatenating segment labels and IDs. For example, the name of segment 3 is "CH2 3", made as a concatenation of the label (CH2) and ID (3).

The same order of atoms must be used in $PDB and $FMOXYZ. The coordinates of atoms in $PDB are ignored. Only ATOM and HETATM records are read. $PDB does not affect fragments in any way, it is solely for segments. If the user wants to look at the interactions of, say, a carboxyl in a side chain of a residue, by changing the segment ID column to a new, unused number, the carboxyl can be assigned to a new segment.

The results are shown in Figure 33. PA is a post-processing for FMO, so in an output, first FMO results are printed, followed by PA. In PA, one obtains the charges of each segment Q_i (the terminal CH_3

```
The best FMO energy is              -15.594561724

...

Segment properties (a.u.); E=E0+EES+EREP+EDI+Esolv
```

i Name NAT	Q_i	q_i	E_i^0	E_i^{ES}	E_i^{REP}	E_i^{DI}	ΔE_i^{solv}	E_i'
1 CH3_1 4	-0.0171	0.0158	-2.8080	0.0003	-0.0431	-0.0009	0.0014	-2.8502
2 CH2_2 3	0.0147	-0.0142	-2.4410	0.0003	-0.0299	-0.0005	0.0002	-2.4708
3 CH2_3 3	0.0024	-0.0023	-2.4395	0.0003	-0.0304	-0.0005	0.0002	-2.4698
4 CH2_4 3	0.0024	-0.0023	-2.4395	0.0003	-0.0304	-0.0005	0.0002	-2.4699
5 CH2_5 3	0.0147	-0.0142	-2.4410	0.0003	-0.0299	-0.0005	0.0002	-2.4708
6 CH3_6 4	-0.0171	0.0157	-2.8080	0.0003	-0.0431	-0.0009	0.0014	-2.8502

```
...

Segment pair properties (kcal/mol); E=EES+EDI+(Esolv=Ees+Enones)
```

i j	R,Å[a]	R,rel[a]	ΔE_{ij}^{ES}	ΔE_{ij}^{DI}	ΔE_{ij}^{es}	ΔE_{ij}^{non-es}	ΔE_{ij}
2 1	1.52	0.51	-0.148	-1.040	0.152	0.000	-1.036
3 1	2.54	0.85	0.067	-0.755	-0.022	0.000	-0.711
3 2	1.53	0.51	-0.151	-0.811	0.124	0.000	-0.838
4 1	3.90	1.30	0.017	-0.171	0.016	0.000	-0.138
4 2	2.54	0.85	0.086	-0.689	-0.065	0.000	-0.668
4 3	1.53	0.51	-0.118	-0.810	0.115	0.000	-0.813
5 1	5.08	1.69	-0.005	-0.050	0.014	0.000	-0.041
5 2	3.91	1.30	-0.023	-0.156	0.026	0.000	-0.152
5 3	2.54	0.85	0.086	-0.690	-0.065	0.000	-0.669
5 4	1.53	0.51	-0.151	-0.811	0.124	0.000	-0.839
6 1	6.40	2.13	0.063	-0.014	-0.053	0.000	-0.003
6 2	5.08	1.69	-0.005	-0.050	0.014	0.000	-0.041
6 3	3.90	1.30	0.017	-0.171	0.016	0.000	-0.138
6 4	2.54	0.85	0.067	-0.755	-0.022	0.000	-0.711
6 5	1.52	0.51	-0.148	-1.040	0.152	0.000	-1.036

```
...

        TOTAL ENERGY (2)=    -15.5945704693[b]
```

[a] For each segment pair, two distances are printed, one in Å, and another is unitless.
[b] The total energy in PA includes a higher order CT, so the energy differs from that in FMO.

Figure 33. Enhanced output for the FMO2-DFTB3/D3(BJ)/SMD/PA input in Table 104.

groups are negative, the inner CH_2 groups are positive), induced solvent charges q_i, interactions ΔE_{ij} between CH_2 and CH_3 groups, all of which turn out to be attractions, and their various components. In SMD, there is no non-es term in PIEs. By looking at the components, it is clear that because solute-solute electrostatics is screened in water, the origin of the attraction is the dispersion (one could criticize this run by pointing out that *n*-hexane is not soluble in water, but a molecule could still be brought in water mechanically).

6.3.7 *FMO2-MP2/PCM, PIEDA + SA*

In $(H_2O)_4$, three waters are arbitrarily assigned to subsystem *A*, and the remaining water to subsystem *B*. In the simplest case of neglecting deformation energies, three calculations are needed to do a subsystem analysis. First, the input file for the complex *AB* is discussed.

The input header is given in Table 106 and the coordinates should be added from Table 98. Analyses are usually done with RUNTYP = ENERGY. RMP2 is selected in $CONTRL. The basis set is defined in $BASIS; because it has d-functions, ISPHER = 1 is used.

Table 106. Water tetramer, FMO2-MP2/PCM/6-31G**, PIEDA + SA (complex AB).

```
$CONTRL NPRINT = −5 RUNTYP = ENERGY SCFTYP = RHF MPLEVL = 2
   ISPHER = 1 $END
$SYSTEM MWORDS = 100 MEMDDI = 100 $END
$GDDI NGROUP = 1 $END
$SCF DIRSCF = .T. NPUNCH = 0 $END
$BASIS GBASIS = N31 NGAUSS = 6 NDFUNC = 1 NPFUNC = 1 $END
$PCM SOLVNT = WATER IEF = −10 ICOMP = 0 ICAV = 1 IDISP = 1 IFMO = −1
   MODPAR = 73 $END
$FMO NBODY = 2 NLAYER = 1 NFRAG = 4 NACUT = 3
MODMOL = 8 MOLFRG(1) = 1,1,1,2 $END
$FMOPRP NPRINT = 9 MODPAR = 8205 IPIEDA = 1 $END
$DATA
DFTB3/3OB E = −16.3313609921 GMAX = 0.0000625 GRMS = 0.0000231
C1
H-1 1
O-1 8
 $END
```

268   *Complete Guide to the Fragment Molecular Orbital Method in GAMESS*

Monomer surface areas (in A**2), charges (a.u.) and solute-solvent energies (kcal/mol).

I	cover,%	q_I	ε_I	d_x^{solv}	d_y^{solv}	d_z^{solv}	ΔE_I^{es}	ΔE_I^{cav}	ΔE_I^{disp}	ΔE_I^{rep}	ΔE_I^{solv} [a]
1	26.0	-0.0003	0.0	-0.109	-0.501	1.341	-5.323	3.863	-2.930	0.604	-3.785
2	25.0	0.0009	0.0	-0.521	0.026	-1.305	-5.312	4.106	-2.645	0.537	-3.314
3	23.4	-0.0037	0.0	0.044	0.581	1.321	-5.233	4.086	-2.679	0.545	-3.280
4	25.6	-0.0006	0.0	0.599	-0.102	-1.317	-5.196	3.882	-2.779	0.567	-3.525

Fragment-wise subsystem partition summary.

I	Name	Sys	E_I^{part}	EBB	ΔE_{Ij}^{U},	$j{=}1,2$ kcal/mol
1	frg00001	1	-76.226548328	-76.217027050	-11.949	**-9.853**[b]
2	frg00002	1	-76.231955856	-76.216247199	-19.715	**-2.208**
3	frg00003	1	-76.225822677	-76.216207312	-12.067	**-9.887**
4	frg00004	2	-76.216611765	-76.216611765	-21.949	0.000

Subsystem partition summary.

I	Charge	E_I'	E_I^{part}	EBB	ΔE_{ij}^{U}, $j{=}1,2$ kcal/mol
1	0.0001	-228.6843268	-228.7018156	-228.6843268	0.000 -21.949
2	-0.0001	-76.2166117	-76.2341005	-76.2166117	-21.949 0.000

[a] Surface areas were truncated.
[b] The same values as PIEs because B has 1 fragment.

Figure 34. Enhanced output for the PIEDA + SA input in Table 106 for the complex *AB*.

MODPAR = 73 in \$PCM and MODPAR = 8205 in \$FMOPRP are set to define the partial screening in PCM. SA is turned on by MODMOL = 8 and the division of fragments into subsystems is set in MOLFRG.

The results are shown in Figure 34. ΔE_{Ij}^{U} (Eq. (57)) for $I \in A$ and $j = 2$ is the same as PIE ΔE_{IJ}, because there is just one unconnected pair (only one fragment J in subsystem $j = B$).

Next, to calculate subsystem A, its coordinates (the first 12 atoms) should be extracted, and the input file adjusted for 3 fragments. The obtained input file is shown in Table 107, where the differences to complex AB are highlighted. In this input, INDAT and \$FMOBND are not used; in general, they have to be adjusted (recreated) for A or B.

When either subsystem (here, B) consists of just a single fragment, its calculation can be done as a non-FMO run or as FMO with 1 fragment (see Section 5.6 for peculiarities of the latter). Here, an input is made as a non-FMO calculation, shown in Table 108. IFMO

Table 107. Water tetramer, FMO2-MP2/PCM/6-31G**, PIEDA + SA (isolated subsystem *A*).

```
$CONTRL NPRINT = -5 RUNTYP = ENERGY SCFTYP = RHF MPLEVL = 2
  ISPHER = 1 $END
$SYSTEM MWORDS = 100 MEMDDI = 100 $END
$GDDI NGROUP = 1 $END
$SCF DIRSCF = .T. NPUNCH = 0 $END
$BASIS GBASIS = N31 NGAUSS = 6 NDFUNC = 1 NPFUNC = 1 $END
$PCM SOLVNT = WATER IEF = -10 ICOMP = 0 ICAV = 1 IDISP = 1 IFMO = -1
  MODPAR = 73 $END
$FMO NBODY = 2 NLAYER = 1 NFRAG = 3 NACUT = 3
MODMOL = 8 MOLFRG(1) = 1,1,1 $END
$FMOPRP NPRINT = 9 MODPAR = 8205 IPIEDA = 1 $END
$DATA
DFTB3/3OB E = -16.3313609921 GMAX = 0.0000625 GRMS = 0.0000231
C1
H-1 1
O-1 8
 $END
$FMOXYZᵃ
```

ᵃAtoms 1–12 (fragments 1–3) should be taken from Table 98.

Table 108. Water tetramer, MP2/PCM/6-31G**, for PIEDA + SA (isolated B).

```
$CONTRL NPRINT = -5 RUNTYP = ENERGY SCFTYP = RHF MPLEVL = 2
  ISPHER = 1 $END
$SYSTEM MWORDS = 100 MEMDDI = 100 $END
SCF DIRSCF = .T. NPUNCH = 0 $END
$BASIS GBASIS = N31 NGAUSS = 6 NDFUNC = 1 NPFUNC = 1 $END
$PCM SOLVNT = WATER IEF = -10 ICOMP = 0 ICAV = 1 IDISP = 1 MODPAR = 3
  $END
$DATAᵃ
DFTB3/3OB E = -16.3313609921 GMAX = 0.0000625 GRMS = 0.0000231
C1
O 8.0  -0.1168848748  -2.9135794830  0.2028922197
H 1.0  -0.2461973358  -1.9570082413  0.0214612021
H 1.0  -0.7115311992  -3.1937861376  0.9035061053
 $END
```

ᵃAtoms 13–16 (fragment 4) are taken from Table 98.

is removed from $PCM. MODPAR = 3 in $PCM is a combination of 1 (parallelization) and 2 (omit gas-phase run). Non-FMO runs (with some exceptions) cannot use GDDI ($GDDI is removed from the input). The partial screening model cannot be used, so the suboption 8 in MODPAR of $PCM is not used. Atomic coordinates for a non-FMO run are put in $DATA.

Alternatively, to run B as FMO with 1 fragment, NBODY = 1 and MOLFRG(1) = 1 should be used, with IPIEDA = 1 removed (PIEDA cannot be done for 1 fragment).

After all calculations of AB, A, and B are done, it is necessary to process the results. Here, the decomposition in Eq. (51) is demonstrated (neglecting the two deformation terms). The collected results are in Table 109.

Table 109.   Water tetramer, FMO2-MP2/PCM/6-31G**, raw data for SA.[a]

Quantity	Source	$I = 1$	$I = 2$	$I = 3$	$I = 4$
$E_I^{\mathrm{part,AB}}$	AB^{b}	−76.22654833	−76.23195586	−76.22582268	−76.21661177
$E_I^{\mathrm{part,X}}$	A,B^{b}	−76.22877357	−76.23150810	−76.22736281	−76.22394511
$\Delta E_I^{\mathrm{part}}$	Eq. (47)	1.396	−0.281	0.966	4.602
$\Delta E_I^{\mathrm{solv,AB}}$	AB^{c}	−3.785	−3.314	−3.280	−3.525
$\Delta E_I^{\mathrm{solv,X}}$	A,B^{c}	−3.959	−3.251	−4.026	−5.985
$\Delta\Delta E_I^{\mathrm{solv}}$	Eq. (49)	0.174	−0.063	0.746	2.460
$\Delta E_I^{\mathrm{pPLd}}$	Eq. (50)	1.222	−0.218	0.220	2.142
ΔE_{IJ}	AB^{d}	−9.853	−2.208	−9.887	

[a]Fragments $I = 1\ldots3$ are in subsystem A, fragment $I = 4$ is in subsystem B. X denotes subsystem, $X = A$ or B. The units for $E_I^{\mathrm{part,X}}$ and $E_I^{\mathrm{part,AB}}$ are hartree, and the rest of energies are in kcal/mol.
[b]For A or AB (fragmented in FMO), $E_I^{\mathrm{part,X}}$ is taken as Epart from "Fragment-wise subsystem partition summary". For B (non-FMO), it is the value in "TOTAL FREE ENERGY IN SOLVENT".
[c]For A or AB (FMO), $\Delta E_I^{\mathrm{solv,X}}$ is taken as Gsol in the output section labeled "Monomer surface areas...", for B (non-FMO), as the value in "TOTAL INTERACTION".
[d]They are PIEs ΔE_{IJ}, for $I \in A$ ($I = 1\ldots3$) and $J \in B$ ($J = 4$), printed as "total" in two-body FMO properties.

Table 110. Water tetramer, FMO2-MP2/PCM/6-31G**, final results in SA.[a]

Quantity	Symbol	$I = 1$	$I = 2$	$I = 3$	$I = 4$
desolvation	$\Delta\Delta E_I^{solv}$	0.174	−0.063	0.746	2.460
polarization[b]	ΔE_I^{pPtd}	1.222	−0.218	0.220	2.142
interaction	$\Delta E_{I,J}{}^{c}$	−9.853	−2.208	−9.887	
binding[d]	ΔE_I^{bind}	−8.457	−2.489	−8.921	4.602

[a]Fragments $I = 1...3$ are in subsystem *A*, fragment $I = 4$ is in subsystem *B*.
[b]The destabilization component of polarization.
[c]$J = 4$ (subsystem *B*).
[d]The sum of the desolvation, polarization, and interaction energies.

The final SA results are in Table 110, where the definition of fragment binding energies in Eqs. (53) and (54) is used. The total binding energy is the sum of four ΔE_I^{bind} values, −15.264 kcal/mol. It is useful to do a check sum, and recompute the binding energy from the total FMO energies according to Eq. (42). These energies can be obtained by extracting the best FMO energy (Section 6.4) from the three output files and computing the binding energy as −304.935916288− (−228.687644485−76.223945116) in hartree, which is −15.265 kcal/ mol, whereas summing the SA binding energies in Table 110 gives the same result (within a round-off error). For converting energies, 1 hartree = 627.509469 kcal/mol can be used.

6.3.8 *FMO3-SCS-MP2/MCP, EDA3*

The input header is given in Table 111 and the coordinates should be added from Table 98. Here, a MCP is used in the easiest for the user way — by specifying PP = MCP and choosing a built-in basis set GBASIS = MCP-DZP (double-ζ with polarization). Hydrogen atoms use no MCP, which is handled internally. The information on how many electrons are treated as particles and how many are replaced by the potential is printed in the output file. Note that this input has no detached bonds for atoms with an MCP basis set, otherwise one would have to generate HMOs (Section 4.11.1).

Table 111.   Water tetramer, FMO3-SCS-MP2/MCP/PCM, EDA3.

$CONTRL NPRINT = –5 RUNTYP = ENERGY SCFTYP = RHF MPLEVL = 2
ISPHER = 1 PP = MCP $END
$SYSTEM MWORDS = 100 MEMDDI = 100 $END
$GDDI NGROUP = 1 $END
$SCF DIRSCF = .T. NPUNCH = 0 $END
$BASIS GBASIS = MCP-DZP $END
$MP2 SCSPT = SCS $END
$PCM SOLVNT = WATER IEF = –10 ICOMP = 0 ICAV = 1 IDISP = 1 IFMO = –1
MODPAR = 73 $END
$FMO NBODY = 3 NLAYER = 1 NFRAG = 4 NACUT = 3 $END
$FMOPRP NPRINT = 9 MODPAR = 8205 IPIEDA = 1 $END
$DATA
DFTB3/3OB E = –16.3313609921 GMAX = 0.0000625 GRMS = 0.0000231
C1
H-1 1
O-1 8
$END

SCS-MP2 is chosen with SCSPT = SCS and MPLEVL = 2. The partial screening model is selected with MODPAR = 73 and MODPAR = 8205. EDA3 is chosen as NBODY = 3 and IPIEDA = 1.

The results are shown in Figure 35, with the focus on three-body related terms. In this SCS-MP2 calculation, the energy was computed without setting MP2PRP = .T., thus dipole moments are for RHF. Three-body properties are contracted according to Eq. (18).

6.3.9 *FMO2-CAM-B3LYP/6-311G*, Density + MEP*

Molecular electrostatic potential plotted on a density isosurface can guide a search of docking poses of a ligand binding to a protein. The input header is given in Table 112 and the coordinates should be added from Table 98. The basis set is specified in $BASIS with a matching ISPHER = 1. It is a triple-ζ basis with polarization, approaching the limit of a basis set that can be used without special measures (Section 5.3); however, the properties of interest are the electron density and MEP, and they are more forgiving than the energy.

Three-body FMO properties.

I	J	K	DL	R_{IJK}^{min}	R_{IJK}^{max}	$\Delta E'_{IJK}$	ΔE^V_{IJK}	ΔE_{IJK}	ΔE^{EX}_{IJK}	ΔE^{CT+MIX}_{IJK}	ΔE^{RC+DI}_{IJK}	ΔE^{solv}_{IJK}
3	2	1	C1	0.67	0.67	-0.891	1.523	0.637	-0.159	0.631	0.159	0.005
4	2	1	C1	0.67	0.67	-0.890	1.519	0.633	-0.159	0.630	0.158	0.004
4	3	1	C1	0.67	0.67	-0.889	1.522	0.637	-0.159	0.633	0.159	0.004
4	3	2	C1	0.67	0.67	-0.889	1.520	0.633	-0.159	0.632	0.158	0.002

Total energy of the molecule: Ecorr(3)= $E^{\text{FMO3-SCS-MP2}}$=-68.631046363

Dipole moment D(xyz),DA(3)= $\mathbf{d}^{\text{FMO3-RHF}}$ = 0.00030 -0.00119 -0.02143 0.02147

Contracted two-body FMO properties (with FMO3 corrections).

I	J	DL	Z	R_{IJ}	$\Delta Q_{I \to J}$	$\Delta \tilde{E}_{IJ}$	ΔE^{ES}_{IJ}	$\Delta \tilde{E}^{EX}_{IJ}$	$\Delta \tilde{E}^{CT+MIX}_{IJ}$	$\Delta \tilde{E}^{RC+DI}_{IJ}$	$\Delta \tilde{E}^{solv}_{IJ}$	a
2	1	C1	0	0.67	-0.0491	-8.694	-16.661	13.483	-4.487	-2.215	1.185	
3	1	C1	0	1.28	-0.0000	-1.919	-2.459	-0.037	-0.346	-0.318	1.241	
3	2	C1	0	0.67	-0.0493	-8.820	-16.720	13.548	-4.504	-2.222	1.078	
4	1	C1	0	0.67	0.0492	-8.755	-16.705	13.549	-4.513	-2.220	1.135	
4	2	C1	0	1.28	-0.0000	-1.974	-2.474	-0.036	-0.347	-0.320	1.203	
4	3	C1	0	0.67	-0.0490	-8.797	-16.673	13.483	-4.475	-2.215	1.083	

Total energy of the molecule: Ecorr(2)= $E^{\text{FMO3-SCS-MP2}}$=-68.631046363 b

[a] Some part of the output was truncated.
[b] FMO3 energy expressed according to the contracted FMO2 form in Eq. (20).

Figure 35. Enhanced output for the FMO3-MP2/EDA input in Table 111.

Table 112. Water tetramer, FMO2-CAM-B3LYP/6-311G*, density + MEP on a grid.

```
$CONTRL NPRINT = −5 RUNTYP = ENERGY SCFTYP = RHF DFTTYP = CAMB3LYP
ISPHER = 1 $END
$SYSTEM MWORDS = 100 MEMDDI = 100 $END
$GDDI NGROUP = 1 $END
$SCF DIRSCF = .T. NPUNCH = 0 SOSCF = .T. DIIS = .F. SHIFT = .T. DAMP = .T. $END
$BASIS GBASIS = N311 NGAUSS = 6 NDFUNC = 1 $END
$GRID SIZE = 0.1 XVEC(1) = 1 YVEC(2) = 1 ZVEC(3) = 1 $END
$FMO NBODY = 2 NLAYER = 1 NFRAG = 4 NACUT = 3 $END
$FMOPRP NPRINT = 9 MODPRP = 564 GRDPAD = 2.0 $END
$DATA
DFTB3/3OB E = −16.3313609921 GMAX = 0.0000625 GRMS = 0.0000231
C1
H.1 1
O.1 8
$END
```

Although this system can converge with the default methods, in order to demonstrate the "levers" to be used in case of bad convergence, SOSCF = .T. DIIS = .F. SHIFT = .T. DAMP = .T. are added to $SCF.

The cube box chosen to encompass the whole molecule is automatically determined, which is accomplished by providing unit vectors in $GRID, and adding 16 to MODPRP in $FMOPRP. Its value 564 breaks into 4 + 16 + 32 + 512, a combination of 4 suboptions. 4 turns on grid computation of the electron density, 16 generates a grid box based on the padding GRDPAD, 32 adds MEP computation, and 512 is used to store grid date in distributed memory. The grid quality is chosen with SIZE = 0.1 Å, which is relatively high.

The results of this calculation are shown in Figure 36. Cube data are written to DAT file as $CUBE groups. For a good grid, "Total density count:" (40.51435) should be close to "Total number of electrons:" (40). Using Eq. (79), one obtains a 1.3% error. Most electron loss occurs because of the grid quality, but some is due to the padding (GRDPAD) that may be too short, so increasing the grid quality may be done in parallel with a larger padding (the more diffuse the basis set, the larger the padding should be).

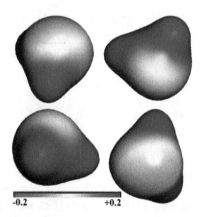

Figure 36. MEP plotted on an isosurface of density $(\rho|(\mathbf{r})| = 0.05)$ for the $(H_2O)_4$ input in Table 112. Pinkish heads are due to positive charges of oxygens, and bluish ears are due to positive charges on hydrogens.

6.3.10 FMO-TDDFT/4-31G, FRET

The way to read in a basis set is demonstrated here (the basis set was taken from the database[11]). The input header is given in Table 113 and the coordinates should be added from Table 98. The functional, LC-BOP (long-range corrected Becke (B) exchange and one parameter

Table 113. Water tetramer, FMO-TDDFT/4-31G, FRET.

```
$CONTRL NPRINT = −5 RUNTYP = ENERGY DFTTYP = LCBOP
TDDFT = EXCITE $END
$SYSTEM MWORDS = 200 MEMDDI = 100 $END
$GDDI NGROUP = 1 $END
$SCF DIRSCF = .T. NPUNCH = 0 $END
$FMO NBODY = 2 NLAYER = 1 NFRAG = 4 NACUT = 3 IEXCIT(1) = −1,1,2,0
IACTFG(1) = 1,0,1,1 $END
$TDDFT MULT = 1 NSTATE = 3 $END
$FMOPRP NPRINT = 9 $END
$DATA
DFTB3/3OB E = −16.3313609921 GMAX = 0.0000625 GRMS = 0.0000231
C1
H 1.0
S  3
1    0.1873113696E + 02    0.3349460434E-01
2    0.2825394365E + 01    0.2347269535E + 00
3    0.6401216923E + 00    0.8137573261E + 00
S  1
1    0.1612777588E + 00    1.0000000
O 8.0
S  4
1    0.8832728600E + 03    0.1755062799E-01
2    0.1331292800E + 03    0.1228292230E + 00
3    0.2990640790E + 02    0.4348835838E + 00
4    0.7978677160E + 01    0.5600108038E + 00
L  3
1    0.1619444664E + 02   −0.1134010029E + 00    0.6854527471E-01
2    0.3780086022E + 01   −0.1772864659E + 00    0.3312254350E + 00
3    0.1070983575E + 01    0.1150407929E + 01    0.7346078781E + 00
L  1
1    0.2838798407E + 00    0.1000000000E + 01    0.1000000000E + 01
$END
```

(OP) correlation), is specified in DFTTYP. TDDFT is selected with TDDFT = EXCITE.

```
States chosen for FRET are marked with a '*'.
FRAGMENT STATE EXCITATION  TRANSITION DIPOLE, A.U.  OSCILLATOR
Fragment State   ωᶠᵢ, eV      X       Y       Z        STRENGTH
        1   1 *    8.056    0.0967 -0.1111 -0.1768     0.010
        1   2     10.017    0.3539  0.3466 -0.2259     0.073
        1   3     10.041   -0.0768 -0.0150 -0.0371     0.002
        3   1 *    8.055    0.0862 -0.1018  0.1874     0.010
        3   2     10.017    0.3406  0.3580  0.2282     0.073
        3   3     10.040   -0.0802 -0.0147  0.0325     0.002
        4   1 *    8.050    0.1000  0.0953  0.1835     0.010
        4   2     10.015   -0.3597  0.3501  0.2069     0.072
        4   3     10.034    0.0244 -0.0873  0.0311     0.002

The FRET matrix Hᶠᴿᴱᵀ is (hartree)
            1           2           3
  1    0.2960592
  2   -0.0001948    0.2960275
  3   -0.0001434    0.0001436    0.2958242
Eigenvalues of the FRET matrix
 STATE  EXCITATION  TRANSITION DIPOLE, A.U.  OSCILLATOR
           eV        X       Y       Z        STRENGTH
  1      8.048    0.0878  0.0942  0.0714      0.004
  2      8.050    0.1355 -0.1440  0.0212      0.008
  3      8.063   -0.0269 -0.0467 -0.3074      0.019
```

Figure 37.   Enhanced output for the FMO-TDDFT/FRET input in Table 113.

In this calculation, three water molecules #1, #3, and #4 (chosen arbitrarily, without any rational reason) were selected to be coupled, and one water #2 was a polarizing bystander, selected with IACTFG(1) = 1,0,1,1. For each chromophore, the excited state number is specified in IACTFG; in this case, it is the first (lowest) excited state (the multiplicity is the same as for the ground state, that is, a singlet). The number of states (the same for each monomer) is specified in NSTATE of $TDDFT. It should not be smaller than the largest number in IACTFG, but it may be larger, for a user inspection of other excited states.

A FRET calculation is a special FMO2 run, so NBODY = 2 should be used. Dimer calculations in FRET are for excitonic couplings only (no DFT or TDDFT calculations are done for dimers), and TDDFT calculations are done for monomers only, so FRET involves FMO1-TDDFT with a coupling calculation for dimers. FRET is selected with IEXCIT(1) = −1,1,2,0.

The results are shown in Figure 37. Among the three excited states for each chromophore fragment, the first one is selected and coupled in the FRET Hamiltonian. The excitonic coupling increased the splitting between chromophore energies, from 0.006 eV to 0.015 eV.

6.3.11 *FMO2-HF3c Optimization*

The input header is given in Table 114 and the coordinates should be added from Table 97. For this small system, RUNTYP = OPTIMIZE would probably be more efficient, but OPTFMO is chosen for a demonstration of the FMO engine. MODGRD = 42 is set to get the exact gradient.

HF-3c is RHF with a certain built-in basis set that should not be changed, and three corrections (3c) (see Section 4.1.2).

To accelerate the geometry optimization and reduce the output file, MODIO = 3072 is used. This is a superoption, that is, it translates into multiple other options set internally, relieving the user from having to learn many technical options. There are two side effects for MODIO: (1) MEM10 is set to 10 MW, which may not suffice. In this case, an error message is printed (it may happen for very large fragments, and if it did, MEM10 in $SYSTEM should be set to a larger value). (2) NFGD in $OPTFMO should be set to the same value as NFRAG in $FMO.

The results of the geometry optimization are shown in Figure 38. The final section of the output shows the last structure, its energy and gradient. An optimization may not find a minimum within the maximum number of steps NSTEP, in which case the final gradient (GMAX and GRMS values) will be larger than the threshold OPTTOL. A

Table 114. *n*-hexane, FMO2-HF3c, optimization.

```
$CONTRL NPRINT = -5 RUNTYP = OPTFMO SCFTYP = RHF $END
$SYSTEM MWORDS = 100 MEMDDI = 100 MODIO = 3072 $END
$GDDI NGROUP = 1 $END
$SCF DIRSCF = .T. NPUNCH = 0 $END
$BASIS GBASIS = HF-3C $END
$FMO NBODY = 2 NLAYER = 1 NFRAG = 3 MODGRD = 42
INDAT(1) = 0 1 -2 7 -11 0 3 -4 12 -15 0 5 -6 16 -20 0 $END
$OPTFMO NSTEP = 200 OPTTOL = 1E-4 NFGD = 3 $END
$FMOBND
-2 3 HF-3C
-5 4 HF-3C
$END
$FMOHYB
HF-3C 5 5
1 0 -0.109774    0.515059    0.000000    0.000000    0.864504
0 1 -0.109774    0.515059    0.815063    0.000000   -0.288167
0 1 -0.109774    0.515059   -0.407531   -0.705865   -0.288168
0 1 -0.109774    0.515059   -0.407531    0.705865   -0.288168
0 1  0.996474    0.015610    0.000000    0.000000    0.000000
$END
$DATA
DFTB3/3OB E = -15.5992436830 GMAX = 0.0000744 GRMS = 0.0000230
C1
H-1 1
C-1 6
$END
```

Using "grep NSERCH:" on the output file	
NSERCH: 1 E=	-233.874633977 GMAX=0.0263782 GRMS=0.0090970
NSERCH: 2 E=	-233.876681814 GMAX=0.0046769 GRMS=0.0014220
...	
NSERCH: 24 E=	-233.877165224 GMAX=0.0000341 GRMS=0.0000123
Located minimum in the output file	

```
NSERCH:24 E=-233.877165224 GMAX=0.0000341 GRMS=0.0000123
Cartesian coordinates for the CONVERGED geometry (Angst.):
C  6  -0.009088742647  -1.478103261037   1.355074063440
C  6  -0.005345487459  -0.033678220141   0.793742645092
...
```

Figure 38. Output for the FMO-HF3c input in Table 114.

geometry optimization may be easily restarted using the last geometry from a previous run.

Checking the progress while a geometry optimization is running (Section 6.4) may be useful to see if something inappropriate may be happening, such as the energy rising or oscillating. The former might happen for 1–2 steps quite commonly but a steady increase may be a sign of serious problems, such as an inappropriate fragmentation or option setup. The latter may happen if MODGRD = 42 was not set (some QM methods cannot use it, see Section 2.6.2.9). There may be other technical issues, such as the grid quality in DFT, tessellation in PCM, etc.

Some energy surfaces are flat, in which case Hessian-based methods (which are the default in OPTIMIZE and OPTFMO) may not be efficient. In this case, one can try gradient-based methods, such as conjugated gradient (CG) in OPTFMO.

6.3.12 *FMO2-RHF/STO-3G:B3LYP/6-31G*/FDD, TS Search*

Before running a transition state search, it is necessary to prepare a guess structure that should be reasonably close to the transition state (with some half-broken chemical bonds stabilized by the environment). For such structure a Hessian is calculated. The input file for a Hessian can be made by simply changing RUNTYP = SADPOINT into RUNTYP = FMOHESS. An analytic Hessian calculation may be expensive. An alternative (not used here) is to use PAVE with a partial semi-numerical Hessian, written out as a full Hessian matrix (Section 3.9).

In a preliminary Hessian run, the user should check the imaginary normal modes. Ideally, only one imaginary frequency is large, and its eigenvector matches the desired reaction. Normal modes can be inspected either by looking at the elements with large weights in the output, or by visualizing the vibration in MacMolPlt or other programs. The vibrational mode should correspond to the desired breaking or forming of chemical bonds. If no such imaginary mode is found, then a different initial structure should be made. If there are several imaginary frequencies, one of which is the desired mode, one can try a transition state search; however, if the frequency of the desired mode is not the largest, it may be necessary to guide the search by changing input options such as IFOLOW (*sic*, 6-character limit) in $STATPT.

For *n*-hexane, the initial structure is not a suitable candidate for a transition state. NSTEP = 1 in $STATPT is used to stop the run just after 2 points (the counting is from 0 like floors in some countries). Normally, NSTEP should be large, such as 100.

The input header is given in Table 115 and the coordinates should be added from Table 97. This is a fairly complex input because a two-layer FDD is used, the lower layer is RHF/STO-3G and the higher layer is B3LYP/6-31G*. The layer-specific selection is in DFTTYP(1) = NONE,B3LYP (RHF and DFT for the lower and higher layer, respectively). The assignment of fragments to layers is done in LAYER(1) = 1,2,2, which says that fragment 1 is in layer 1, and fragments 2 and 3 are in layer 2.

The general plan of this run is to optimize fragment 3. However, atom 5 is the BDA for the bond connecting the optimized and frozen areas, so it was decided to exclude atom 5 from the list of optimized atoms, defined in IACTAT(1) (see Figure 27). The active domain **A** is specified in IACTFG, in this case, a single fragment 3. Frozen **F** and polarizable **B** domains are specified in LAYER(1) as layers 1 and 2, respectively (**A** is a part of **B**). In this system, the optimized fragment 3 forms a covalent bond to fragment 2. It is required for the FD method that all such fragments be included in **B**, therefore, the minimal **B** should include fragments 2 and 3. Fragment 2 corresponds to domain **b** (see Figure 17). Note that normally, domain **B** should include all fragments within the recommended threshold (Section 4.7) from **A**, that is, **B** is usually much larger than just 2 fragments. Here, a small **B** is used for demonstration.

FDD (FD with an extra dimer approximation) is chosen with MODFD = 3. Without a non-zero MODFD, the input would be a regular multilayer calculation. MODGRD = 42 is needed to get the accurate gradient. The basis sets are chosen in $DATA, with the layer indicated after a minus sign, so that "H-2" defines the basis set for hydrogen in layer 2. 6-31G* uses no polarization for hydrogen (6-31G), but for C, the polarization function should be specified

Table 115.   *n*-hexane, FMO2-RHF/STO-3G:B3LYP/6-31G*/FDD, TS search.

$CONTRL NPRINT = –5 RUNTYP = SADPOINT SCFTYP = RHF DFTTYP =
  B3LYP $END
$SYSTEM MWORDS = 100 MEMDDI = 100 $END
$GDDI NGROUP = 1 $END
$SCF DIRSCF = .T. NPUNCH = 0 $END
$STATPT OPTTOL = 1E-4 NSTEP = 1 IACTAT(1) = 6 16 –20 $END
$FMO NBODY = 2 NLAYER = 2 NFRAG = 3 MODGRD = 42
DFTTYP(1) = NONE,B3LYP LAYER(1) = 1,2,2 MODFD = 3 IACTFG(1) = 3
INDAT(1) = 0 1 –2 7 –11 0 3 –4 12 –15 0 5 –6 16 –20 0 $END
$FMOPRP NPRINT = 9 $END
$FMOBND
–2 3 STO-3G 6-31G*
–5 4 STO-3G 6-31G*
$END
$FMOHYB ᵃ
STO-3G 5 5

...

6-31G* 15 5

...

$END
$DATA
DFTB3/3OB E = –15.5992436830 GMAX = 0.0000744 GRMS = 0.0000230
C1
H-1 1
STO 3

C-1 6
STO 3

H-2 1
N31 6

C-2 6
N31 6
D 1 ; 1 0.8 1.0

$END
$HESS
ENERGY IS   -157.3549102988 E(NUC) IS   155.0327246348
 1 1 1.00000000E-08 0.00000000E + 00 0.00000000E + 00 0.00000000E + 00
   0.00000000E + 00
...ᵃ

ᵃTruncated.

```
(FMOHESS run)
ANALYZING SYMMETRY OF NORMAL MODES...
    FREQUENCIES IN CM**-1, IR INTENSITIES IN DEBYE**2/AMU-ANGSTROM**2,
    REDUCED MASSES IN AMU.
                      1         2         3         4         5
      FREQUENCY:  143.75 I    0.15      0.15      0.15      0.15

(SADPOINT run):
        HESSIAN MODE FOLLOWING SWITCHED ON, FOLLOWING MODE   1
        FOLLOWING MODE  1 WITH EIGENVALUE   -0.00058 AND COMPONENTS
...
NSERCH: 0 E= -157.3549102991 GRAD. MAX= 0.0098733 R.M.S.= 0.0041144
NSERCH: 1 E= -157.3553124944 GRAD. MAX= 0.0026036 R.M.S.= 0.0013005
...
        OVERLAP OF CURRENT MODE  18 WITH PREVIOUS MODE IS      0.835611
  WARNING!! MODE SWITCHING WAS FOLLOWING MODE 1, NOW FOLLOWING MODE 18
        FOLLOWING MODE 18 WITH EIGENVALUE    0.00082 AND COMPONENTS
```

Figure 39. Output for the FMO/FDD input in Table 115 and a preceding FMOHESS calculation.

explicitly (from the database[11]). The Hessian matrix, taken from the DAT file of the preliminary run, is read as $HESS.

The results are shown in Figure 39. For reference, an excerpt from the Hessian run is shown, with the frequency of the imaginary mode ($143.75i$ cm^{-1}). This small frequency is suggestive that the initial structure is not suitable for finding a transition state.

The fiasco in the output of RUNTYP = SADPOINT is indeed clear to a trained eye. On step 1, a normal mode switching of mode 1 to 18 occurred. This means that after the Hessian had been updated using the gradient of step 0, the followed normal mode of this updated Hessian became root 18, whereas normally it should have stayed root 1 all along the search. What is even more lacrimoniously dolorous is that the frequency became real (not imaginary, $\sqrt{0.00082}$ a.u.). Among the normal modes of the updated Hessian, the mode with the largest overlap to the original normal mode (assumed to be the search direction) is selected. A transition state search is a tricky operation, which requires a vigilant check of its progress.

6.3.13 *FMO2-LC-BOP/ 6-31G, PAVE*

The input header is given in Table 116 and the coordinates should be added from Table 98. In the partial Hessian run using PAVE, segments are used, selected with MODPAN = 33 (1 + 32), where 1 turns on PA and 32 defines segments to be equal to fragments, so INDATP and/or $PDB are not needed. For PA, RUNTYP should be FMOHESS and MODGRD = 42 is used for an accurate gradient (in this semi-numerical run, analytic gradients are numerically differentiated). It should be noted that analytic Hessian may also be used in PAVE with MODGRD = 10 and METHOD = ANALYTIC in $FORCE.

The two main parameters of PAVE are: NZPHA = 6 defines that 6 lowest (possibly, imaginary) modes should be skipped in the calculation of thermodynamical properties, and JACTAT(1) = 7 −12 defines that atoms from 7 to 12 (that is, segments 3 and 4) are active, so the Hessian is computed with respect to their coordinates. It is not necessary to assign whole segments as active atoms, but it is sensible to do so.

Table 116. Water tetramer, FMO2-LC-BOP/D3(BJ)/6-31G, PAVE, IR + Raman spectra.

```
$CONTRL NPRINT = –5 RUNTYP = FMOHESS SCFTYP = RHF DFTTYP =
   LCBOP $END
$SYSTEM MWORDS = 100 MEMDDI = 100 MODIO = 3072 $END
$GDDI NGROUP = 1 $END
$SCF DIRSCF = .T. NPUNCH = 0 $END
$DFT DC = .T. IDCVER = 4 $END
$BASIS GBASIS = N31 NGAUSS = 6 $END
$FORCE METHOD = SEMINUM PROJCT = .F.
NZPHA = 6 JACTAT(1) = 7 –12 VIBSIZ(1) = 1E-3,2E-3 $END
$FMO NBODY = 2 NLAYER = 1 NFRAG = 4 NACUT = 3 MODGRD = 42 $END
$FMOPRP MODPAN = 33 $END
$DATA
DFTB3/3OB E = –16.3313609921 GMAX = 0.0000625 GRMS = 0.0000231
C1
H-1 1
O-1 8
   $END
```

The first parameter 10^{-3} in VIBSIZ(1) defines the shift of coordinates (in bohr) for numerical derivatives (it is not recommended to use values much larger or much smaller than 5×10^{-3}). The second parameter 2×10^{-3} defines the shift for the electrostatic field numerical derivatives. When it is not zero, 19 additional gradient calculations are performed to get Raman activities. IR intensities are automatically computed in any PAVE run.

RDFT is chosen with SCFTYP = RHF DFTTYP = LCBOP, where LC-BOP is a long range-corrected (LC) DFT with the Becke (B) exchange and one parameter (OP) correlation. The basis set is 6-31G. MODIO = 3072 is used to reduce the output, because in this run many single point calculations are performed.

The results of this calculation are shown in Figure 40. Thermodynamical quantities are meaningful for a stationary point (for which the gradient is 0). For the geometry used in this calculation, the gradient is far from 0, and there are three imaginary frequencies.

```
Imaginary vibrational frequencies in cm-1 are:
   392.195  338.940  153.029
 Real vibrational frequencies in cm-1 are:
   58.292  96.743 128.853 158.667 177.174 249.713 326.383 752.046
   893.113 1622.353 1649.402 3525.400 3574.306 3956.682 3958.570
Segment thermochemistry in kcal/mol at  298.15 K and 0.10132E+06 Pa
```

$$G_i = E_i^{ZPE} + H_i^T - TS_i \quad \text{(Svib is in cal/mol/K)}$$

i	E_i^{ZPE}	H_i^T	$-TS_i$	G_i	S_i
1	0.000	0.000	0.000	0.000	0.000
2	0.000	0.000	0.000	0.000	0.000
3	14.921	0.779	-1.373	14.327	4.604
4	14.877	0.636	-1.114	14.399	3.737
total	29.798	1.415	-2.487	28.726	8.341

MODE	FREQ(CM**-1)	RED. MASS	IR INTENS.	RAMAN ACT.	DEPOLARIZ.[a]
1	392.195	1.036712	3.729484	4.141607	0.749790
2	338.940	1.051507	2.680827	10.588280	0.707928

...

[a] This Table provides discrete IR and Raman spectra.

Figure 40.   Enhanced output for the PAVE input in Table 116.

Only two segments have non-zero vibrational energies, because they include active atoms. IR intensities and Raman activities are given in the table at the end of Figure 40, for each frequency.

6.3.14 *FMO2-DFTB3/PBC, NVT US MD, FA*

The input header is given in Table 117 and the coordinates should be added from Table 97. The MD simulation is chosen with RUNTYP = MD. In $MD, DT is the time step (1 fs). NSTEPS = 200 is the number of steps. BATHT sets the temperature (298 K). MDINT = VVERLET NVTNH = 2 MBT = .TRUE. MBR = .TRUE. RSTEMP = .T. define an

Table 117. *n*-hexane, FMO2-DFTB3/PBC, NVT US MD, FA.

```
$CONTRL NPRINT = −5 RUNTYP = MD $END
$SYSTEM MWORDS = 100 MEMDDI = 100 MODIO = 3072 $END
$GDDI NGROUP = 1 $END
$SCF DIRSCF = .T. NPUNCH = 0 $END
$BASIS GBASIS = DFTB $END
$DFTB SCC = .T. DAMPXH = .T. DFTB3 = .TRUE. PARAM = 3OB-3-1 DISP = UFF
  MODGAM = 13
  PBCBOX(1) = 5.0000   0.0000   0.0000
              0.0000  10.0000   0.0000
              0.0000   0.0000  10.0000
$END
$MD
  MDINT = VVERLET DT = 1.0D-15 NVTNH = 2 NSTEPS = 200
  MBT = .TRUE. MBR = .TRUE. BATHT = 298.0 RSTEMP = .T.
  JEVERY = 10 KEVERY = 100 IRATTL(1) = 1,1,1
  TRJFMT = F14O6 MODFLU = 1
  USAMP = .T. RZERO(1) = 1.5 UFORCE(1) = 10 IPAIR(1) = 1,2 IUSTYP = 0
$END
$FMO NBODY = 2 NLAYER = 1 NFRAG = 3 INDAT(1) =  0 1 −2 7 −11 0 3 −4 12
  −15 0 5 −6 16 −20 0
MODGRD = 32 $END
$FMOBND
−2 3 3OB-3-1
−5 4 3OB-3-1
$END
```

(Continued)

Table 117.   (*Continued*)

$FMOHYB
3OB-3-1 4 4
1 0 0.565124 0.000000 0.000000 −0.825006
0 1 0.565123 0.777823 0.000000 0.275002
0 1 0.565124 −0.388911 0.673614 0.275003
0 1 0.565124 −0.388911 −0.673614 0.275003
 $END
 $DATA
DFTB3/3OB E = −15.5992436830 GMAX = 0.0000744 GRMS = 0.0000230
C1
H 1
C 6
 $END

NVT simulation conducted from the scratch (not as a restart). For a restart, $MD should be taken from the restart file.

MD quantities such as the kinetic energy are printed once in JEVERY = 10 steps; the restart and trajectory files will be updated once in KEVERY = 100 steps. RATTLE is turned on by IRATTL(1) = 1,1,1 (the three values turn it on separately for C–H, C–N, and C–O bonds). TRJFMT = F14O6 defines that every coordinate in the trajectory file is written in 14 decimal places with 6 digits after the decimal point (F14.6 format in FORTRAN). MODFLU = 1 turns on the fluctuation analysis.

US is selected with USAMP = .T., with the bond distance (IUSTYP = 0) between atoms 1,2 (IJPAIR) constrained to the value of 1.5 Å (RZERO) with the force constant of 10 kcal mol^{-1} Å^{-2} (UFORCE).

The PBC cell is defined as PBCBOX, rectangular with the dimensions of 5 Å × 10 Å × 10 Å. Atoms are not spilled so NSPILL in $DFTB is not set. MODGAM = 13 is an acceleration option for DFTB. MODIO = 3072 is used to reduce output. DFTB parameters are set in $DFTB (for PBC the only dispersion model is UFF), with the matching set of HMOs in $FMOBND and $FMOHYB, and atoms without layer indices in $DATA. MODGRD = 32 is used to get the accurate gradient.

```
Monomer energies in MD:

I     E_I'^0     ⟨ΔE_I'⟩      Epotmin    Epotmax    Ekin0     <Dekin>

1  -4.803117  0.004398  -4.807020  -4.786462  0.008961   0.001220
2  -5.635127  0.008580  -5.635724  -5.614347  0.008206  -0.000207
3  -4.810330  0.009269  -4.810330  -4.788764  0.011896   0.000101
 Monomer temperatures (=scaled kinetic energies) in MD (K):
I   N_I^at      T0(min)            T_I
1    7     269.50572356      306.21083568
2    6     287.92170647      280.64566725
3    7     357.77542720      360.82535412
 Global temperature (K), T0=    305.92491471 T=    317.65636660
 Dimer potential energies in MD:

I J      ΔE_IJ^0        ⟨ΔΔE_IJ⟩         Emin          Emax

2 1   -0.12820072   -0.00509432   -0.13978501   -0.12820072
3 1   -0.00137417    0.00023549   -0.00146126   -0.00093677
3 2   -0.13503887    0.00227504   -0.13800166   -0.12588038

    QM potential energy   = ⟨E⟩   -15.4935257362
```

Figure 41. Enhanced output for the MD/FA input in Table 117.

The FA results of the MD simulation are shown in Figure 41. Another important outcome is a set of US coordinates, which can be used to get PMF.

6.3.15 *Lego Input Maker*

Above, 14 complete input examples are provided. As an aid for making general input files, a complete input group for each method is listed in Table 118. Inevitably, only the most general usage is covered.

6.4 Processing Results

A small collection of useful Unix commands to process results is provided in Table 119. In addition, the command to return system semaphores (if not enough are available, DDI may fail to start) is

ipcrm -s `ipcs | grep 00000000 | awk '{ print $2 }'`.

Table 118. Typical groups for making input files.[a]

Method	Input
RHF	$CONTRL SCFTYP = RHF $END
HF-3c	$CONTRL GBASIS = HF-3c $END
MP2	$CONTRL MPLEVL = 2 $END
CCSD(T)	$CONTRL CCTYP = CCSD(T) $END
DFT	$CONTRL DFTTYP = CAMB3LYP $END
TDDFT	$CONTRL TDDFT = EXCITE $END
	$FMO IEXCIT(1) = L,2,1,2 $END[b]
DFTB	$BASIS GBASIS = DFTB $END
	$DFTB SCC = .T. DAMPXH = .T. DFTB3 = .TRUE. PARAM = 3OB-3-1 $END
ROHF	$CONTRL SCFTYP = ROHF[c] $END
	$FMO SCFFRG(1) = RHF,...,ROHF[c],...RHF MULT(I) = P $END
UHF	$CONTRL SCFTYP = UHF[c] $END
(UDFT)	$FMO SCFFRG(1) = RHF,...,UHF[c],...RHF MULT(I) = P $END
MCSCF [d]	$CONTRL SCFTYP = MCSCF $END
	$FMO SCFFRG(1) = RHF,...,MCSCF,...RHF IJVEC(1) = I,0,0,1,M MULT(I) = P $END
	$DET NCORE = J NACT = K NELS = L $END[c]
	$VEC1...$END
D3(BJ) [e]	$DFT DC = .T. IDCVER = 4 $END
PCM	$PCM SOLVNT = WATER IEF = −10 ICOMP = 0 ICAV = 1 IDISP = 1 IFMO = −1 $END
SMD	$PCM SOLVNT = WATER IEF = −10 SMD = .T. IFMO = −1 $END
PA	$FMOPRP MODPAN = 33 $END
PIEDA	$FMOPRP IPIEDA = 1 $END
PA,PIEDA	$PCM MODPAR = 73 $END
	$FMOPRP MODPAR = 8205 $END

[a]If a method is listed several times, all keywords should be used.
[b]L is the number of the chromophore fragment.
[c]Some of the fragments should be set as ROHF (UHF). The multiplicity for ROHF(UHF) fragments I is set as P.
[d]There should be 1 MCSCF fragment I and the rest is RHF. I, J, K, L, M, and P should be set as appropriate. $VEC1 should contain M initial RHF molecular orbitals of the MCSCF fragment.
[e]To be used in HF, DFT, or DFTB.

Table 119. Collection of data processing commands, applied to an FMO output file $job.log.[a]

FMO task	Command
Total energy.	grep "The best FMO energy" $job.log
Optimization report for OPTIMIZE, OPTFMO, or SADPOINT.	grep NSERCH: $job.log
Reaction path for IRC.	grep -A 1 "DISTANCE STOTAL" $job.log
List of US coordinates.	grep "US bond" $job.log[b]

[a]Can be applied in real time while a job is still running.
[b]May have to be adjusted depending on the type of the coordinate.

6.5 GAMElish Dictionary

You may be reading a GAMESS manual and seeing a phrase that does not make sense, despite being worded in English, which may even be your mother tongue (even worse if it is not). A word may look like a typical English word, perhaps with a Latin prefix, a Germanic root, or a Norman suffix, but its meaning may still elude you. Try to use this dictionary for translations from GAMElish into English. Some of the items are loan words from other languages.

[input] deck
  A named group of input options, such as $CONTRL.

dictionary [file]
  A binary file $job.F10 (with rank extensions in GDDI) containing run-time data such as atomic coordinates, MOs, Fock matrix, etc. In rare cases, F10 can be used for restarts.

distributed [array]
  An array allocated in shared memory by data servers.

flag
  A logical option taking the value .TRUE. or .FALSE.

group
> A set of compute processes doing some task together, with "assistant" data servers. Group sizes refer only to compute processes.

master
> The first compute process in each group is its master (with the local rank of 0). For GDDI, there is one master per group, one grandmaster per world, and one demiurge for the universe. Any master does computations like all other members of the group (a master is not a dedicated process). The master node is the first node in the list of nodes.

megaword
> One million of words (q.v.), abbreviated as MWord or MW.

processor
> In most instances, it means a CPU core. A GAMElish message about a memory amount *on each processor* means "on each CPU core" in English.

punch [file]
> A text file (called DAT in this book) with restart data such as MOs, Hessians, etc., mostly written in formatted groups. Punch can be a verb (*the results are punched* in GAMElish can be translated as "the results are written to the DAT file"). A hint to the possible etymology of the term may be found on the front cover.

rank
> ID for a parallel process, with the lowest rank of 0. Any compute process or data server in GAMESS has a unique rank; the lower half is for compute processes, the upper half is for data servers. When processes write data to rank-specific files, the rank is used as the last extension (usually, a 3-digit number), as in $job.F06.005 for rank 5.

replicated [array]
> An array separately allocated by each compute process.

scratch [directory]
> A directory for run-time files, set as SCR in rungms.

species
> In DFTB, it roughly means a chemical element, so "3 species" typically correspond to 3 different chemical elements (in a rare usage, there may be multiple species for the same element).

word
> An amount of memory to store one floating point number of double precision. Nowadays, 1 word is commonly equal to 8 bytes.

6.6 Suggestions for Further Reading

Topic	Reference
Detailed equations in FMO	T. Nagata, D. G. Fedorov, K. Kitaura (2011). Mathematical formulation of the fragment molecular orbital method. In R. Zalesny, M. G. Papadopoulos, P. G. Mezey, J. Leszczynski (Eds.), *Linear-Scaling Techniques in Computational Chemistry and Physics*, Springer, pp. 17–64.
GDDI	D. G. Fedorov, R. M. Olson, K. Kitaura, M. S. Gordon, S. Koseki (2004). A new hierarchical parallelization scheme: generalized distributed data interface (GDDI), and an application to the fragment molecular orbital method (FMO). *J. Comput. Chem.* **25**, 872–880.
PIEDA	D. G. Fedorov, K. Kitaura (2007). Pair interaction energy decomposition analysis. *J. Comput. Chem.* **28**, 222–237.
FEDA on S_N2 reactions	D. G. Fedorov, T. Nakamura (2022). Free energy decomposition analysis based on the fragment molecular orbital method. *J. Phys. Chem. Lett.* **13**, 1596–1601.
SA	D. G. Fedorov, K. Kitaura (2016). Subsystem analysis for the fragment molecular orbital method and its application to protein-ligand binding in solution. *J. Phys. Chem. A* **120**, 2218–2231.
PA	D. G. Fedorov (2020). Partition analysis for density-functional tight-binding. *J. Phys. Chem. A* **124**, 10346–10358.
Recent FMO research by various groups	A. Heifetz, Ed. (2020). *Quantum Mechanics in Drug Discovery.* Springer. Y. Mochizuki, S. Tanaka, K. Fukuzawa, Eds. (2021). *Recent Advances of the Fragment Molecular Orbital Method.* Springer.

(*Continued*)

(*Continued*)

Topic	Reference
Nanomaterial fragmentations (zeolite and silicon nanowire)	D. G. Fedorov, J. H. Jensen, R. C. Deka, K. Kitaura (2008). Covalent bond fragmentation suitable to describe solids in the fragment molecular orbital method. *J. Phys. Chem. A* **112**, 11808–11816.
	D. G. Fedorov, P. V. Avramov, J. H. Jensen, K. Kitaura (2009). Analytic gradient for the adaptive frozen orbital bond detachment in the fragment molecular orbital method. *Chem. Phys. Lett.* **477**, 169–175.
1 million atom nanomaterials	Y. Nishimoto, D. G. Fedorov, S. Irle (2014). Density-functional tight-binding combined with the fragment molecular orbital method. *J. Chem. Theory Comput.* **10**, 4801–4812.
	Y. Nishimoto, D. G. Fedorov (2018). Adaptive frozen orbital treatment for the fragment molecular orbital method combined with density-functional tight-binding. *J. Chem. Phys.* **148**, 064115.
Protein crystal optimization	T. Nakamura, T. Yokaichiya, D. G. Fedorov (2021). Quantum-mechanical structure optimization of protein crystals and analysis of interactions in periodic systems. *J. Phys. Chem. Lett.* **12**, 8757–8762.
Crystal surface adsorption	T. Nakamura, T. Yokaichiya, D. G. Fedorov (2022). Analysis of guest adsorption on crystal surfaces based on the fragment molecular orbital method. *J. Phys. Chem. A* **126**, 957–969.
Solid state catalysis	T. Nakamura, D. G. Fedorov (2022). The catalytic activity and adsorption in faujasite and ZSM-5 zeolites: the role of differential stabilization and charge delocalization. *Phys. Chem. Chem. Phys.* **24**, 7739–7747.

Instructions for Accessing Online Supplementary Material

The online supplementary material for this book includes a collection of input and output files; most of these files are for Section 6.3, but a few input files used in other Sections are also provided. Readers are encouraged to download the sample files to facilitate their comprehension of the accompanying discussion in the book.

1. Register an account/login at https://www.worldscientific.com
2. Go to: https://www.worldscientific.com/r/13063-SUPP
3. Download the Supplementary Material from https://www.worldscientific.com/worldscibooks/10.1142/13063#t = suppl

For subsequent download, simply log in with the same login details in order to access.

References

1. M. W. Schmidt, K. K. Baldridge, J. A. Boatz, S. T. Elbert, M. S. Gordon, J. H. Jensen, S. Koseki, N. Matsunaga, K. A. Nguyen, S. Su, T. L. Windus, M. Dupuis, J. A. Montgomery (1993). General atomic and molecular electronic structure system. *J. Comput. Chem.* **14**, 1347–1363.
2. G. M. J. Barca, C. Bertoni, L. Carrington, D. Datta, N. De Silva, J. E. Deustua, D. G. Fedorov, J. R. Gour, A. O. Gunina, E. Guidez, T. Harville, S. Irle, J. Ivanic, K. Kowalski, S. S. Leang, H. Li, W. Li, J. J. Lutz, I. Magoulas, J. Mato, V. Mironov, H. Nakata, B. Q. Pham, P. Piecuch, D. Poole, S. R. Pruitt, A. P. Rendell, L. B. Roskop, K. Ruedenberg, T. Sattasathuchana, M. W. Schmidt, J. Shen, L. Slipchenko, M. Sosonkina, V. Sundriyal, A. Tiwari, J. L. Galvez Vallejo, B. Westheimer, M. Włoch, P. Xu, F. Zahariev, M. S. Gordon (2020). Recent developments in the general atomic and molecular electronic structure system. *J. Chem. Phys.* **152**, 154102.
3. M. S. Gordon, D. G. Fedorov, S. R. Pruitt, L. V. Slipchenko (2012). Fragmentation methods: a route to accurate calculations on large systems. *Chem. Rev.* **112**, 632–672.
4. ABINIT-MP. https://ma.issp.u-tokyo.ac.jp/en/app/92.
5. PAICS. http://www.paics.net/index_e.html.
6. K. Fukuzawa, S. Tanaka (2022). Fragment molecular orbital calculations for biomolecules. *Curr. Opin. Struct. Biol.* **72**, 127–134.
7. D. G. Fedorov (2017). The fragment molecular orbital method: theoretical development, implementation in GAMESS, and applications. *Wiley Interdiscip. Rev. Comput. Mol. Sci.* **7**, e1322.

8. D. G. Fedorov (2021). Recent development of the fragment molecular orbital method in GAMESS. In Y. Mochizuki, S. Tanaka, K. Fukuzawa (Eds.), *Recent Advances of the Fragment Molecular Orbital Method*, Springer, pp. 31–51.
9. GAMESS homepage. https://www.msg.chem.iastate.edu/gamess/download.html.
10. G. D. Fletcher, M. W. Schmid, B. M. Bode, M. S. Gordon (2000). The distributed data interface in GAMESS. *Comput. Phys. Commun.* **128**, 190–200.
11. Basis set exchange. https://www.basissetexchange.org/.
12. VMD 1.9.3. https://www.ks.uiuc.edu/Research/vmd/.
13. Facio 23.1.5. http://zzzfelis.sakura.ne.jp/.
14. FU 4.0. https://sourceforge.net/projects/fusuite/.
15. MacMolPlt 7.7.2. https://brettbode.github.io/wxmacmolplt/downloads.html.
16. Mercury 4.2.0. https://www.ccdc.cam.ac.uk/community/csd-community/freemercury/.
17. PDB2PQR 3.5.1. https://server.poissonboltzmann.org/pdb2pqr.
18. PyMOL. https://pymol.org/2/.
19. pyProGA 1.2. https://gitlab.com/Vlado_S/pyproga.
20. K. Kitaura, K. Morokuma (1976). A new energy decomposition scheme for molecular interactions within the Hartree-Fock approximation. *Int. J. Quantum Chem.* **10**, 325–340.
21. K. Murata, D. G. Fedorov, I. Nakanishi, K. Kitaura (2009). Cluster hydration model for binding energy calculations of protein-ligand complexes. *J. Phys. Chem. B* **113**, 809–817.
22. Grossfield. WHAM: the weighted histogram analysis method, version 2.0.10. http://membrane.urmc.rochester.edu/content/wham/.
23. The DFTB website. http://dftb.org.

Index

Author's Note: Page in bold is the main page to look at, for entries with multiple pages listed.

$AFOMOD, 57
$BASIS, 34, 137
$CCINP, 172
$CIS, 152
$CONTRL
    CCTYP, 40, **148**
    CITYP, 40, **152**
    CONV, 172
    DFTTYP, 39
    EXETYP, **22**, 148
    ICUT, 204
    ISPHER, **34**, 255
    ITOL, 204
    LOCAL, **179**, 184
    MAXIT, 172
    MPLEVL, 146
    NPRINT, 160
    NUMCOR, 146
    NZVAR, 113
    PP, 141
    QMTTOL, 204
    RUNTYP, **11**, 12
    TDDFT, 152
$DET, 149
    CVGTOL, 172

ITERMX, 172
NACT, 149
NCORE, 149
NELS, 149
NSTATE, 149
PURES, 149
WSTATE, 149
$DFT
    DC, 135
    IDCVER, 135
    NLEB, 214
    NRAD, 214
    SG1, 213
    SWOFF, 134
$DFTB
    CPCONV, 172
    DAMPXH, 136
    DISP, 136
    EMU, 136
    ISPDMP, 138
    ITYPMX, 169
    LATOPT, 111
    LCDFTB, 136
    MAXINT, 113
    MODESD, 217

MODGAM, **138**, 213
MXCPIT, 172
NDFTB, 136
PARAM, 137
PBCB, 138
SCC, 136
$DFTBAO, 137
$DISREP
  DKA, 71
  RWA, 71
$ECP, 141
$EFRAG, 68
$ELDEN, 164
$ELMOM
  IEMOM, 131
  WHERE, 131
$ELPOT, **131**, 163, 164
$FMO, 37
  ATCHRG, 203
  CCTYP, 40
  DFTTYP, 40
  EXFID, 152
  FRGNAM, 41
  IACTFG, 158
  ICHARG, 39
  IEXCIT, 154
  INDAT, 38
  INDATP, 38
  LAYER, 40
  MAXBND, 49
  MAXCAO, **49**, 181
  MAXKND, 49
  MAXRIJ, 217
  MODCHA, 131
  MODGRD, 47
  MODMOL, 42
  MODMUL, 214
  MOLFRG, **42**, 104
  MPLEVL, 40

MULT, 39
NACUT, 38
NATCHA, 203
NBODY, 38
NDUALB, 205
NFRAG, 37
NFRND, 159
NLAYER, 40
NOPFRG, 40
NOPSEG, **40**, 42
RAFO, 184
RCORSD, 43
RESDIM, 43
RITRIM, 43
SCFFRG, 39
SCFTYP, 40
SCREEN, 201
TDDFT, 40
$FMOBND, 53
$FMOEFP, 200
  IEABDY, 200
  IEACAL, 200
  IPEFP, 200
  IPFMO, 200
  ITRLVL, 200
  NLEVEL, 68
  NPRIEA, 200
$FMOHYB, 56
$FMOPRP
  CNVPCM, 52
  CONV, 52
  COROFF, **52**, 134, 209
  E0BDA, **50**, 194
  EFMO0, **50**, 194
  EINT, 194
  EINT0, 50
  EPLODS, **50**, 194
  GRDPAD, **50**, 163
  IJVEC, **54**, 207

IPIEDA, **50**, 195
IREST, **54**, 208
LOADBF, **54**, 66
LOADGR, **54**, 66
MANNOD, 21, **54**, 64, 65
MAXAOC, 150, **195**
MAXIT, 52
MCONFG, 170
MCONV, **52**, 170
MODCHA, 213
MODORB, 54
MODPAN, **50**, 97
MODPAR, **53**, 217
MODPRP, **51**, 163, 208
MOFOCK, 50
N0BDA, 50
NAODIR, 212
NBUFF, 217
NCVSCF, **52**, 170
NGRFMO, 21, **54**, 66
NGRID, **50**, 164
NGUESS, **51**, 169
NLCMO, **50**, 161
NPCMIT, **52**, 73
NPRINT, **53**, 210, 211, 217
PRTDST, 54
R0BDA, 50
VDWRAD, 43
$FMOXYZ, 37
$FORCE
   JACTAT, 99
   KDIAGH, 212
   METHOD, 121
   NPRHSS, 121
   NZPHA, 99
   PURIFY, 121
   SCLFAC, 122
   TEMP, 99, **122**, 209
   VIBSIZ, **122**, 124

$GDDI
   MANNOD, 21
   NGROUP, **21**, 225
   NSUBGR, 24, **225**
   PAROUT, 211
$GLOBOP, 68
$GRID
   ORIGIN, 163
   SIZE, 163
   UNITS, 163
   XVEC, 163
   YVEC, 163
   ZVEC, 163
$GUESS, 150
$HESS, **123**, 124
$INTGRL, 212
$IRC
   FORWRD, 125
   NPOINT, 125
$MASS, 121
$MCP, 142
$MCSCF, 149, 172
$MD
   BATHT, 118
   CCMS, 119
   CFORCE, 119
   DROFF, 119
   DT, 118
   IPAIR, 118
   IRATTL, 118
   IUSTYP, 118
   JEVERY, 107, **118**
   KEVERY, **119**, 206
   MODFLU, 118
   NATCC, 119
   NSTEPS, 118
   NVTNH, 118
   RZERO, 118
   SFORCE, 119

SSBP, 119
TRJFMT, 58, **118**
UFORCE, 118
USAMP, 118
$MEX
　IMEXFG, 129, **130**
　MULT2, 130
　NSTEP, 172
　SCF2, 130
　TGMAX, 172
　TGRMS, 172
$MP2
　CODE, 146
　MP2PRP, 132
　SCSPT, 148
$OPTFMO
　FHMCON, 116
　IACTAT, 116
　IFREEZ, 115
　IHMCON, 116
　IREST, **116**, 206
　MAXNAT, 116
　METHOD, **115**, 217
　NFGD, **116**, 245
　NPRICO, 116
　NSTEP, 113, 115
　OPTTOL, 113, **115**
　SHMCON, 116
$OPTRST, 206
$PCM
　EPS, 69
　EPSINF, 156
　ICAV, 71
　ICOMP, 71
　IDISP, 71
　IEF, 71
　MODPAR, **74**, 78, 213
　MXSP, 75
　MXTS, 71

NESFP, 75
SMD, 70
SOLVNT, 71
TABS, 210
$PCMCAV
　ALPHA, 72
　RADII, **71**, 72
　RIN, **71**, 76
　XE, 76
　YE, 76
　ZE, 76
$PCMITR, 172
$PDB, **87**, 117
$PDC, 131
$RAMAN, 124
$SCF
　DAMP, **168**, 171
　DIIS, **167**, 171
　DIRSCF, **211**, 212, 171
　DIRTHR, 168
　FDIFF, **168**, 171
　LOCOPT, **168**, 171
　NPREO, 160
　NPUNCH, 207, **211**
　RESET, 168
　RSTRCT, 166
　SHIFT, **168**, 171
　SOSCF, **167**, 171
　SWDIIS, 170
$STATPT
　DRMAX, 114
　FHMCON, 115
　IACTAT, **116**, 159
　IFOLOW, 280
　IFREEZ, 116
　IHMCON, 115
　NSTEP, 113
　OPTTOL, 113
　SHMCON, 115

$STONE, 131
$SYSTEM
    BALTYP, 63
    KDIAG, 212
    MEM10, 211
    MEM22, 211
    MEMDDI, **22**, 147, 164
    MODIO, 37, 59, 117, **210**,
        211, 217
    MWORDS, 22
    PARALL, 24
$TDDFT
    CNVTOL, 172
    IROOT, 152
    NLEB, 213
    NONEQ, 156
    NRAD, 213
    NSTATE, 276
    TDPRP, 152
$TESCAV
    MTHALL, 74
    NTSALL, 74
$VEC, 150, **207**, 211, 259
$ZMAT, 113
0 state, 79

AFO *see* Covalent fragment
    boundary
Allosteric regulation, 102
Analyses, 84
    EDA *see* Energy decomposition
        analysis
    EDA*n see* Interaction
    FA *see* Fluctuation analysis
    FEDA *see* Free energy
        decomposition analysis
    IEA *see* Interaction, *see*
        Interaction
    PA *see* Partition analysis

PAVE *see* Partition analysis of
    vibrational energy
PIEDA *see* Interaction
SA *see* Subsystem analysis
AO *see* Atomic orbitals
AP *see* Auxiliary polarization
ASC *see* Polarizable continuum
    model
Atomic charge
    ESP fitting, 131
    fixed, 202
    in segments, **85**, 95
    Mulliken population, 130
    printing, **53**, 131
    Stone analysis, 131
Atomic orbitals *see* Basis set
    (definition)
Atomic charges, 130
Auxiliary polarization (AP), 205
    example, 257

BAA *see* Covalent fragment
    boundary
Backbone (BB)
    energy, 83
    splitting in residues, 40
    splitting in segments, 88
Basis set
    auxiliary in RI, 147
    definition, 34
    diffuse, 204
    Dunning, 36
    energy correction, 134
    energy correction (BS), 91
    exchange (database), 36
    for ECP, 142
    for MCP, 142
    in DFTB, 137
    input errors, 245

linear dependence, 204
multiple, **35**, **156**, 197, 205
Pople, 36
read from external file, 36
superposition error *see* Basis set
  superposition error
Basis set superposition error
  (BSSE), 204
  counterpoise correction (CP),
    91
  geometric counterpoise
    correction (GCP), 134
  short-range basis set
    incompleteness (SRB), 134
BB *see* Backbone
BDA *see* Covalent fragment
  boundary
Binding, 78
  active domains, 157
  active site, 35, **157**
  analysis *see* Subsystem
    analysis
  desolvation *see* Solvent effects
  difference to interaction, 81,
    82
  energy, 100
  free vibrational energy, 99
  per fragment, 104
  protein-ligand, **79**, 81, 84, 99
BS *see* Interaction
BSSE *see* Basis set superposition
  error

CC *see* Coupled cluster
CCSD *see* Coupled cluster
CCSD(T) *see* Coupled cluster
CG *see* Geometry optimization
Charge instability, 165
Charge mixer, 168

Charge transfer (CT)
  between fragments, 9
  energy, 8, **90**
  in trimers, 9
  state, 80
Charge transport, 253
CI *see* Configuration interaction
CIS *see* Configuration interaction
Configuration interaction (CI), 40
  with singles (CIS), 152
Configuration state functions (CSF),
  149
Connected dimer, 83
CONV *see* $FMOPRP, *see*
  $CONTRL
Coordination number, 119
COROFF *see* $FMOPRP
Coupled cluster (CC), 148
  approximations, **45**, 46
  completely renormalized (CR-
    CC(2,3)), 148
  core correlation problem, 146
  restricted open-shell (ROCC),
    148
  with singles and doubles
    (CCSD), 148
  with singles, doubles and
    perturbative triples
    (CCSD(T)), 148
Covalent fragment boundary, 172
  adaptive frozen orbitals (AFO),
    183
  bond attached atom (BAA), 55,
    173, 183
  bond detached atom (BDA),
    55, 85, **173**, 183, 185
  hybrid orbital projection
    (HOP), 175
  PIE correction, 185

CP *see* Basis set superposition error
CPHF *see* Hartree-Fock
CR-CC(2,3) *see* Coupled cluster
Crystals *see* Periodic boundary conditions
Crystals
    interactions between fragments *see* Interaction
    catalysis in solid state, 292
    covalent, 191
    molecular, **192**, 203, 292
    nano, 215
    protein, **83**, 292
    quasi-periodic embedding, 203
    surface adsorption, 292
CSF *see* Configuration state functions
CT *see* Charge transfer
CT state, 79
Cygwin, 28
D2 *see* Dispersion
D3 *see* Dispersion

DAT, 14, **58**
Dative bonds, 86
DDI *see* Parallelization
Density functional theory (DFT), 133
    long-range corrected (LC-DFT), 133
    restricted (RDFT), 133
    time-dependent (TDDFT), 151
    unrestricted time-dependent (UTDDFT), 151
    unrestricted (UDFT), 133
    with dispersion correction (DFT-D), 135

Density-functional tight-binding (DFTB), 135
    long-range corrected (LC-DFTB), 135
    non-charge-consistent (NCC-DFTB), 135
    self-consistent charge (SCC-DFTB), 135
    third order (DFTB), 135
Desolvation *see* Solvent effects
DFT *see* Density functional theory
DFTB *see* Density-functional tight-binding
DFT-D *see* Density functional theory
DI *see* Interaction
DIIS *see* $SCF
Dimer, 5
DIRSCF *see* $SCF
Dispersion *see* Interaction
Dispersion
    ab initio, **146**, 148
    empirical models, **135**, 136
EA *see* Electron affinity
ECP *see* Effective core potential
EDA *see* Energy decomposition analysis
EDA2 *see* Interaction
EDA3 *see* Interaction

Effective core potential (ECP), 140
Effective fragment potential (EFP), 67
EFP *see* Effective fragment potential
Electron affinity (EA), 166
Electrostatic Potential (ESP), 4
Electrostatic potential (ESP) *see* Embedding

Embedding, 200
  approximation inconsistency, 202
  approximations, 44
  atomic charge models, 130
  damping, 202
  electronic as in QM/MM, 203
  electronic in FMO/EFP, 69
  energy contribution, 8
  fixed in PL0, 195
  from solvent, 6
  gradient, 47
  hybrid, 201
  mechanical in FMO/MM, 139
  monomer loop, 4
  multilayer, 156
  none, **194**, 197
  partially fixed, 202
  polarization, 80
  quasi-crystal, 202
  with exchange, 161
Energy decomposition analysis (EDA), 89
Energy minimization *see* Geometry optimization
Enthalpy, 98
Entropy, 98, 110
ES *see* Interaction
ESP *see* Embedding
EX *see* Interaction

F06, **14**, 234
F07, **14**, 234
F08, 234
F09, 234
F10, **59**, 234
F15, 234
F22, 234
F23, 234

F30, **59**, 194
F40, **58**, 195, 208, 234
FA *see* Fluctuation analysis
FD *see* Frozen domain
FDD *see* Frozen domain
Fluctuation analysis (FA), 107
  example, 285
FMO/F *see* Molecular orbitals
FMO/FX *see* Molecular orbitals
FMO
  RESPAP, 43
  RESPPC, 43
FMO0, 38, 154, **195**
FMO1, **5**, 154, 172
FMO2, **5**, 38, 83, 107, 216
FMO3, **6**, 38, 46, 202, 216
Förster resonance electron transfer (FRET), 154
  example, 275
  exciton, 155
  excitonic coupling, 155
  Hamiltonian, 155
  solvent screening, 155
Free energy, 98, **108**
FRET *see* Förster resonance electron transfer
Frozen domain (FD), 157
  with dimers (FDD), 158
GCP *see* Basis set superposition error
GDDI *see* Parallelization
Geometry optimization, 113
  acceleration, 51, **210**
  example, 277
  freezing coordinates, 116
  in internal coordinates, 113
  of lattice, 111
  oscillations, 48
  partial *see* Frozen domain

restart, 206
surface crossing *see* Minimum
   energy crossing
surface crossing, 129
transition state, 113
with conjugated gradient (CG),
   279
with constrained internal
   coordinates, 115
with force fields, 68, **139**
Gradient
   approximations, 47
   availability, 11
   projection, 48
   static electric field, 123
   with respect to lattice vectors,
      111
Grandmasters *see* Parallelization
GVB, 261

Hartree-Fock (HF), 11
   coupled-perturbed (CPHF), 47
   restricted open-shell (ROHF),
      144
   restricted (RHF), 11
   time-dependent (TDHF), 151
   unrestricted (UHF), 144
   with 3 corrections (HF-3c),
      134
Hessian, 120
   analytic, 121
   anharmonicity, 122
   isotopes, 121
   partial analysis (PHA), 121
   purification, 121
   semi-numerical, **121**, 206, 224
   thermochemistry, 122
   use in geometry optimization,
      206

use in IRC, 125
use in transition state search,
   124
with PAVE engine *see* PAVE

HF *see* Hartree-Fock
HF-3c *see* Hartree-Fock
HOMO *see* Molecular orbitals
HOP *see* Covalent fragment
  boundary
IACTAT *see* $OPTFMO, *see*
  $STATPT
IEA *see* Interaction, *see* Interaction
IFMO *see* $PCM
IJVEC *see* $FMOPRP

Interaction, 78
   0-order in DFTB (0), 91
   basis set correction (BS), 91
   charge transfer (CT), 90
   contracted PIE, 11
   conventions (es vs ES etc), 13
   definition of PIE, 7
   difference to binding, 81, **82**
   dispersion (DI), 91
   electrostatic (ES), 90
   energy analysis (IEA), 193, **198**
   example, **93**, 267, 271
   exchange-repulsion (EX), 90
   heat map, 93
   hydrogen bond coupling, 5
   in periodic systems, 82
   many-body analyses (EDA*n*),
      89
   many-body energy (MBIE), 89
   metal-ligand, 86
   mixed (MIX), 90
   of excitons, 8
   pair energy (PIE), 5

PIE decomposition analysis (PIEDA), 89
protein-ligand, 35, **42**
remainder correlation (RC), 91
solvent screening (solv) *see* Solvent effects
total energy (TIE), 10
Intrinsic reaction coordinate (IRC), 125
Ionization potential (IP), 166
IP *see* Ionization potential
IPIEDA *see* $FMOPRP
IRC *see* Intrinsic reaction coordinate

Junction rule *see* Covalent fragment boundary

LC-DFT *see* Density functional theory
LC-DFTB *see* Density-functional tight-binding
LCMO *see* Molecular orbitals
LCMOX *see* Molecular orbitals
Liquids, 192
    ionic, 188
Logical node *see* Parallelization
LUMO *see* Molecular orbitals

MANNOD *see* $FMOPRP
Many-body expansion (MBE)
    contracted, 11
    multipoles, 10
    of atomic charges, **10**, 131
    of electron densities, 162
    of energies, 6
    of Fock matrices, 160
    of solute potentials, 73

Many-body interaction energy (MBIE), 89
Masters *see* Parallelization
Materials
    DNA, 189
    enzymes, 157, 189
    liquid, 188
    nano, 190
    polymers, 189
    proteins, 189
MAXCAO *see* $FMO
MAXIT *see* $CONTRL, *see* $FMOPRP
MBE *see* Many-body expansion
MBIE *see* Many-body interaction energy
MCP *see* Model core potential
MCSCF *see* Multiconfigurational self-consistent field
MD *see* Molecular dynamics
MECP *see* Minimum energy crossing
MEMDDI *see* $SYSTEM
Memory, 22
    distributed (MEMDDI), 22
    electronic disk, 212
    for storing integrals, 145, **212**
    replicated (MWORDS), 22
    system pool, 22
    System V, 19
    usage in CC, 148
    usage in MP2, 147
    usage reduction, **23**, 216
MEP *see* Molecular electrostatic potential
Metallicity, 165
MEX *see* Minimum energy crossing
MFMO *see* Multilayer FMO

Minimum energy crossing (MEX), 129
  point (MECP), 129
MIX *see* Interaction
MO *see* Molecular orbitals
Model core potential (MCP), 140
MODGAM *see* $DFTB
MODGRD *see* $FMO
MODIO *see* $SYSTEM
MODPAR *see* $PCM, *see* $FMOPRP
MODPRP *see* $FMOPRP
Molecular dynamics (MD), 117
  chemical reaction, 125
  ensembles, 118
  example, 285
  potential of mean force (PMF), 128
  pre-equilibration, 127
  radial distribution function (RDF), 119
  restart, 206
  trajectory and restart files, 58
  umbrella sampling (US), 125
  window, 126
Molecular electrostatic potential (MEP), 162
  example, 272
  solvent screening, 163
Molecular mechanics
  link atom, 139
  via EFP, 68
  via SIMOMM, 139
Molecular orbitals (MO), 159
  Fock matrix expansion (FMO/F), 160
  highest occupied (HOMO), 161

hybrid for fragment boundaries, 56
LCMO with exchange (LCMOX), 161
LCMOX example, 252
linear combination (LCMO), 160
lowest unoccupied (LUMO), 161
Møller-Plesset perturbation theory, 40, 146
  analytic gradient, 47
  approximations, **45**, 46
  core correlation problem, 146
  density derived properties, 132
  IR intensities, 121
  resolution of the identity (RI-MP2), 147
  restricted open-shell (ROMP2), 144
  restricted (RMP2), 146
  second order (MP2), 40
  spin-component scaled (SCS), 147
  unrestricted (UMP2), 144
Monomer, 4
MP2 *see* Møller-Plesset perturbation theory
MPI *see* Parallelization
Multiconfigurational self-consistent field (MCSCF), 148
  CI basis, 149
  convergers, 169
  example, 259
  orbital conversion, 150
  orbital reordering, 150
  state averaged (SA), 149

Multilayer FMO, 156
MWORDS *see* $SYSTEM
MXATM, 12

NACUT *see* $FMO
NBODY *see* $FMO
NCC *see* Density-functional
  tight-binding
NFRAG *see* $FMO
NGRFMO *see* $FMOPRP
NGROUP *see* $GDDI
NGUESS *see* $FMOPRP
NINTIC *see* $INTGRL
NLAYER *see* $FMO
NOPFRG *see* $FMO
NOPSEG *see* $FMO
NPRINT *see* $FMOPRP, *see*
  $CONTRL
NPUNCH *see* $SCF
NSUBGR *see* $GDDI

Packed option, 32
Parallelization
  compute process, 19
  data server, **19**, 223
  distributed data interface
    (DDI), 19
  efficiency, 219
  file usage, 14
  generalized distributed data
    interface (GDDI), 20
  golden rule, **61**, 232
  grandmasters, 21
  grid computing, 63
  hyperthreading, **20**, 219
  load balancing, **63**, 228
  logical node, **29**, 225
  masters, **20**, 28
  MPI, **24**, 225

network latency, 147
polling problem, 20
sockets, **15**, 19, 24
synchronization, **226**, 229,
  233
three-level GDDI (GDDI/3),
  224
Partition analysis of vibrational
  energy (PAVE), 97
  active atoms, 99
  example, 283
  excluded frequencies, 99
  Hessian in full dimension, 124
  IR spectra, 123
  Raman spectra, 123
  segment definition, 87
Partition analysis (PA), 94
  example, 262
  residue splitting, 41, **88**
  segment definition, 87
PAVE *see* Partition analysis of
  vibrational energy
PBC *see* Periodic boundary
  conditions
PBCBOX *see* $DFTB
PCM *see* Polarizable continuum
  model
Periodic boundary conditions
  (PBC), 111
  atom spilling, 112
  atom wrapping, 112
  cell charge, 190
  lattice optimization, 111
  replication of symmetry-related
    cells, 60
PHA *see* Hessian
PIE *see* Interaction
PIEDA *see* Interaction
PL0 state, 79

PL state, 79
PMF *see* Molecular dynamics
Polarizable continuum model
  (PCM), 6, 69
    apparent solvent charge (ASC),
      69
    cavity, 69
    charge escape, 71
    conductor (C-PCM), 70
    integral equation formalism
      (IEF-PCM), 70
    non-electrostatic terms, 10
    tessera, 69
Polarization, **79**, 101
    by ligand (protein), 81
    by solvent (solute), 9, 110, **198**
    coupling to other interactions,
      196, 200
    destabilization component, 81,
      102, **196**, 197
    in AP *see* Auxiliary
      polarization
    in periodic systems, 195
    in PIEDA, 193
    many-body, 80
    one-body, 80
    stabilization component,
      81, 196, 197
Proteins
    as a medium in PCM, 70
    charge instability, 166
    crystals, 83
    desolvation, 74
    docking, 162
    fragmentation, 84
    in complex with another
      protein, 60
    in complex with ligand, **41**,
      99, 156

photoactive, 156
residue networks (PRN),
  60
PRTDST *see* $FMOPRP

Quantum confinement effect, 165

Radicals, 132, **144**
RAFO *see* $FMO
RC *see* Interaction
RCORSD *see* $FMO
RDF *see* Molecular dynamics
RDFT *see* Density functional
  theory
RESDIM *see* $FMO
Resolution of the identity (RI), 147
RESPPC *see* $FMO
RHF *see* Hartree-Fock
RI *see* Resolution of the identity
RI-MP2 *see* Møller-Plesset
  perturbation theory
RITRIM *see* $FMO
RMP2 *see* Møller-Plesset
  perturbation theory
ROCC *see* Coupled cluster
ROHF *see* Hartree-Fock
ROMP2 *see* Møller-Plesset
  perturbation theory
RST, 58
rungms, 25, 59
    DDI_LOGICAL_NODE_SIZE,
      225
    GMSPATH, 16
    SCR, 17
    USERSCR, 17
RUNTYP *see* $CONTRL

SA *see* MCSCF, *see* Subsystem
  analysis

Salt bridge, 167, **175**
SA-MCSCF *see* MCSCF
SCC *see* Density-functional
  tight-binding
SCF *see* Self-consistent field
SCR *see* rungms
Screening *see* Solvent effects
SCS-MP2 *see* Møller-Plesset
  perturbation theory
SCZV *see* Self-consistent Z-vector
Self-consistent field (SCF), 4
    multiconfigurational *see*
      MCSCF
Self-consistent Z-vector (SCZV), 47
SIMOMM *see* Molecular
  mechanics 108
SMD *see* Solvation model density
SIMOMM *see* Molecular
  mechanics, 139
SO(3), 178
Sockets *see* Parallelization
solv *see* Interaction
Solvation model density (SMD), 69
  non-electrostatic terms, 10
Solvent effects
    charge quenching, 77
    desolvation penalty, 74
    local dielectric constant, 76
    screening, 76
    solvatochromic shift (explicit),
      153
    solvatochromic shift (implicit),
      155
SOSCF *see* $SCF
Spectra
    adsorption, 155
    emission, 155
    infrared (IR), 123
    Raman, 123

Spin
    contamination, 144
    coupling in dimers and
      trimers, 145
    definition, 39
    density, 51, **132**, 162
    surface crossing *see* Minimum
      energy crossing
    transfer, 132
SRB *see* Basis set superposition
  error
Subsystem analysis (SA), 100
    example, 267
    for fragments, 103
    for segments, 105
Sulfur bridges, 182

TDDFT *see* Density functional
  theory
TDHF *see* Hartree-Fock
Temperature, 209
    electronic, 169
    in MD, 107, **118**
    in PCM, 210
    in vibrational energies, 98
    of fragments in MD, 108
Tessera *see* Polarizable continuum
  model
TIE *see* Interaction
Tinker, 138
Transition state (TS), 99
    barrier decomposition analysis,
      110
    example, 279
    in molecular dynamics, 108,
      125
    search, 124
    see Geometry optimization
Trimer, 6

TRJ, 58
TS *see* Transition state

UDFT *see* Density functional
  theory
UFF *see* Dispersion
UHF *see* Hartree-Fock
UMP2 *see* Møller-Plesset
  perturbation theory
Unconnected dimer, 83

US *see* Molecular dynamics
USERSCR *see* rungms
UTDDFT *see* Density functional
  theory

Water bridges, 67

Zero point energy (ZPE),
  98
ZPE *see* Zero point energy

CPSIA information can be obtained
at www.ICGtesting.com
Printed in the USA
BVHW050932090323
659800BV00002B/59

9 789811 263620